About Resources for the Future *and* RFF Press

Resources for the Future (RFF) improves environmental and natural resource policy-making worldwide through independent social science research of the highest caliber. Founded in 1952, RFF pioneered the application of economics as a tool for developing more effective policy about the use and conservation of natural resources. Its scholars continue to employ social science methods to analyze critical issues concerning pollution control, energy policy, land and water use, hazardous waste, climate change, biodiversity, and the environmental challenges of developing countries. RFF Press supports the mission of RFF by publishing book-length works that present a broad range of approaches to the study of natural resources and the environment. Its authors and editors include RFF staff, researchers from the larger academic and policy communities, and journalists.

About ESMAP

The **Energy Sector Management Assistance Program (ESMAP)** is a global technical assistance program that promotes the role of energy for poverty reduction and economic growth. It helps build consensus and provides policy advice on sustainable energy development to governments of developing countries and economies in transition. ESMAP was established in 1983 under the joint sponsorship of the World Bank and the United Nations Development Programme as a partnership with other donors in response to global energy crises. Since its creation, ESMAP has operated in over 100 different countries, through approximately 450 activities covering a broad range of energy issues.

Resources for the Future

To our parents

THE URBAN HOUSEHOLD ENERGY TRANSITION

Social and Environmental Impacts in the Developing World

Douglas F. Barnes

Kerry Krutilla

William F. Hyde

Resources for the Future
Washington, DC, USA

Energy Sector Management Assistance Program
Washington, DC, USA

An RFF Press book
Published by Resources for the Future
1616 P Street NW
Washington, DC 20036–1400
USA
www.rffpress.org

A copublication of Resources for the Future (RFF) and the Energy Sector Management Assistance Program (ESMAP).

Library of Congress Cataloging-in-Publication Data

Barnes, Douglas F.
  The urban household energy transition : social and environmental impacts in the developing world / Douglas F. Barnes, Kerry Krutilla, and William F. Hyde.
    p. cm.
  Includes bibliographical references.
  ISBN 1-933115-06-8 (hardcover : alk. paper) — ISBN 1-933115-07-6 (pbk. : alk. paper)
  1. Dwellings—Energy consumption—Developing countries. 2. Power resources—Developing countries. I. Krutilla, Kerry. II. Hyde, William F. III. Title.
  TJ163.5.D86B35 2004
  333.79'63'091724—dc22
          2004021905

The paper in this book meets the guidelines for permanence and durability of the Committee on Production Guidelines for Book Longevity of the Council on Library Resources. This book was typeset in Minion by Maggie Powell. It was copyedited by Bonnie Nevel. The cover was designed by Maggie Powell. Cover photograph of woman cooking in Bangladesh village, by Prabir Mallik, courtesy of the World Bank. Cover design by Nancy Bratton Design.

ISBN 1-933115-06-8 (cloth) ISBN 1-933115-07-6 (paper)

Contents

Figures and Tables

Preface

One may wonder why we decided to prepare a book on the urban energy transition in developing countries. After all, we all know that people start out using wood in the least developed countries and end up using electricity and other modern fuels in the most developed countries. Such simple characterizations of the energy transition belie the complications in the movement of cities from one stage to another. The complications are important to countries for several reasons. Policies at the macro-level are important because much money is involved in importing the energy that fuels the transition. The stress on the local environments can be triggered by improper policies at some stages of the transition. Finally, at the micro-level, poor people are involved in a day to day struggle to improve their quality of life, and energy is an important way to alleviate many drudgeries associated with tasks such as cooking, lighting, and heating their homes. Thus, by necessity the topics covered in *The Urban Household Energy Transition* span many different disciplines—environmental studies, sociology, economics, and others.

This book is intended to inform policymakers and those interested in development policy in developing countries of the vast array of options that are available for improving quality of life and avoiding environmental problems associated with the household energy transition. We hope the book will assist development specialists who wrestle with problems of resources and poverty, government officials who must make critical decisions concerning the directions of their countries, and academics and authorities involved in the social sciences and environmental aspects of energy and urban development.

At the time we began our study, the information on interfuel substitution and the energy transition was largely anecdotal. People familiar with a single city or town—say, one that had adopted kerosene rapidly when it became available—would describe this town as typical and prescribe policies based on that experience. Studies of individual urban areas were more extensive in scope and more sophisticated in method. Yet they still offered little in the way of useful generalizations or policy guidance because they often described radically different circumstances, with some study cities rapidly switching to modern fuels and other taking decades to make the transition. The purpose of this study was to go beyond anecdotes and reports on individual cities. The method used to transcend the limitations of the individual city studies involved examining the patterns of urban interfuel substitution systematically over a variety of cities, with a variety of energy policies and a variety of local resource conditions.

This book develops a comprehensive assessment of the evolution of residential fuel choice and consumption in urban areas in the developing world, and it examines the effect of urban growth on periurban forest resources. The research is based on a comprehensive analysis of a series of household energy surveys performed under the auspices of the Energy Sector Management Assistance Program (ESMAP) of the World Bank. From 1984 to 2000, this program produced more than 25,000 household energy surveys in 45 cities spanning 12 countries and 3 continents. Additionally, global information system (GIS) mapping software was used to compile a database of site-specific vegetation patterns surrounding a subsample of 34 cities. Taken together, the energy surveys and the biomass data contained sufficiently wide variation in urban fuel choice and consumption patterns, local resource conditions, and energy policy regimes to enable an assessment of the factors underlying the evolution of urban fuel utilization and forest resources. By comparing the patterns of energy use of a large number of cities, we were able to distill a comprehensive picture of both the diversity underlying the energy transition and the fundamental principles applicable across cases.

The preparation of this book was a challenging project, with many complicated overarching issues, including the necessity of assembling and analyzing large and complicated data sets from many different countries. We did so to be confident that the results are not just limited to one region or are the result of some unique regional or urban characteristics. We decided that it would be better to present the results in a comprehensive way, rather than break them up. We hope that we have achieved a broader understanding of the energy transition that will be of assistance in solving real problems and furthering existing research involving urban equity, energy, and the environment in developing countries.

Acknowledgments

This book is the culmination of the work on household energy in developing studies financed and supported by the Energy Sector Management Assistance Program of the World Bank during the last 25 years. We owe much debt and gratitude to the dedicated staff and consultants who carried out household energy surveys and wrote the individual country reports as part of this program. We are especially appreciative of the support of Willem Floor, Robert van der Plas, and Karl Jechoutek, who made significant contributions to the policy analysis in this study. Although numerous authors contributed to this body of work, we would like to particularly thank Philippe Durand, Kevin Fitzgerald, Josephine Arpelange, Rene Masse, Boris Utria, Robert Taylor, and Noureddine Berrah. Paul Wolman also provided valuable advice and assistance for the preparation of the various chapters in the book.

In addition, we would like to acknowledge the contributions of the original research team that shaped the design and conception of the study. The team included Jeffrey Dowd, who worked long and hard on the policy analysis and much of the household data analysis in this study. Also, Liu Qian was invaluable in compiling and standardizing the household energy database, a task that took more than a year; he also helped conduct the data analysis. The biomass estimation and mapping were ably completed by Marnie Tyler, with technical assistance from Paul Ryan and Keith Openshaw. That effort took more than a year. For the case study of Hyderabad, we worked with Mazoor Alam, Jayant Sathaye, Keith Openshaw, Geeta Reddy, and M. Naimuddin; we are grateful for their contributions. Finally, we appreciate the comments and insights of Mark Pitt, who helped with the research design at an early stage in the study. Joy Dunkerley provided a thorough review of an earlier manuscript, and Gerald Foley and Gerald Leach provided their generous insights on the issue assessed in the study. We thank all these individuals for their dedication and support. However, any errors or omissions in the study, as well as the conclusions of the analysis, are our exclusive responsibility.

We also wish to thank Bonnie Nevel for the editing and preparation of the manuscript for publication.

The research was financed by the World Bank. However, the findings, interpretation, and conclusions expressed in this book are entirely our own and should not be attributed in any manner to the World Bank, to its affiliated organizations, or to members of its Board of Executive Directors or the governments they represent.

DOUGLAS F. BARNES
KERRY KRUTILLA
WILLIAM F. HYDE

About the Authors

Douglas F. Barnes is a senior energy specialist within the Energy and Water Department of the World Bank and research scientist in the Department of Sociology at the University of Maryland. Prior to joining the Bank, he worked at the Center for Energy Policy Research at Resources for the Future. He has been involved in social aspects of development for both rural and urban areas for more than 25 years, leading efforts to develop a strategy for rural energy for the World Bank Group, which was published as *Rural Energy and Development: Improving Energy Supplies for Two Billion People.* He is also a leading expert on the socioeconomic impact of rural electrification and is the author of *Electric Power for Rural Growth: How Electricity Affects Rural Life in Developing Countries.*

Kerry Krutilla is an associate professor at the School of Public and Environmental Affairs (SPEA), Indiana University-Bloomington. His current research is in the fields of public choice and environmental policy, growth modeling and sustainable development, and resource policy and the environment, with ongoing research projects studying the systems dynamics in growth models with environmental resources, and game theoretic models of environmental policy with rent-seeking agents. He has taught graduate courses in cost-benefit analysis, managerial economics, and sustainable development, and undergraduate courses in environmental economics and finance. Dr. Krutilla has conducted contract research for a variety of government agencies and organizations, including the U.S. Department of Energy, the Economic Research Service of the U.S. Department of Agriculture, and the World Bank.

William F. Hyde's career in forest economics and policy spans 30 years and includes terms with the U.S. Forest Service, Duke University, and Resources for the Future. He is a senior associate of the Center for International Forestry Research (CIFOR) in Bogor, Indonesia, a visiting professor at the Environmental Economics Unit of Goteborg University in Sweden, and an adjunct professor at the Center for Chinese Agricultural Policy, Chinese Academy of Sciences, Beijing. He has been a consultant to more than two dozen organizations in more than 30 countries. His books include *Timber Supply, Land Allocation and Economic Efficiency, Forestry Sector Intervention* (with Roy Boyd), *Economics of Forestry and Rural Development: Empirical Evidence from Asia* (with Gregory Amacher), and *China's Forests: Global Lessons from Market Reforms* (with Brian Belcher and Jintao Xu).

A Note about Data and Tables

The tables in this volume are based on original data from urban household energy surveys. For each urban area, means for different types of data were calculated for the city and income quintiles within the city.

The income quintiles—rather than the raw survey data—were used to calculate overall values for all cities in the study. Thus, the main dataset for the urban analysis contains information for 45 cities and 235 income groups. Such a method weights each city and each income quintile within a city equally, avoiding biases in which larger cities at the high end of the energy transition outweigh smaller cities in the earlier stages.

Using income class not only offers a degree of standardization but allows us to view the reaction of income classes to different policies and other variables.

Energy Conversion Factors

	Energy content			Efficiency for cooking
Fuel type	Megajoules	Kgoe	Kilo-calories	Percentage
LPG (kg)	45.0	1.059	10,800	60
Electricity (kWh)	3.6	0.085	860	75
Kerosene (liter)	35.0	0.824	8,400	35
Charcoal (kg), 5% Moist. C. 4% Ash	30.0	0.706	7,200	22
Wood (kg), 15% Moist. C. 1% Ash	16.0	0.376	3,840	15
Coal (kg) varies significantly	23.0	0.541	5,520	NA
Dung (kg) 15% Moist. C. 20% Ash	14.5	0.341	3,480	NA
Straw (kg) 5% Moist. C. 4% Ash	13.5	0.318	3,240	NA

Abbreviations and Acronyms

kgoe   kilogram of oil equivalent
kWh   kilowatt hour
kW   kilowatt
km   kilometer
kV   kilovolt
LPG   liquefied petroleum gas
MWh   megawatt hour
MW   megawatt
TWh   terawatt hour

1

Urban Household Energy, Poverty, and the Environment

Traditionally, the words "developing country" evoked pastoral images of rural villages and small-scale agriculture. Following World War II, the dramatic growth of urban populations added scenes of crowded city streets filled with motor scooters, trucks carrying goods and fuels, and vendors cooking food. By 1980, about 900 million people lived in urban areas in the developing world; today, there are more than 2 billion urban dwellers. Rural population growth rates have leveled off in many developing countries, while urban growth rates now average more than 3.3 percent annually. In some cities, urban growth rates have reached levels of 7 to 8 percent per year (World Bank 1998).

The topic of this book is the relationship between urban growth in developing countries and the decisions of urban households to select and consume different kinds and amounts of residential energy. In the earliest stages of a city's development, urban dwellers largely consume biomass-based "traditional" fuels. As a city develops and modernizes, the pattern of residential fuel consumption shifts, often to a succession of transition fuels, such as kerosene or coal, and ultimately to the so called "modern fuels"—liquefied petroleum gas (LPG) and electricity. This book assesses the factors that shape this "urban energy transition," documenting the way in which energy markets in urban areas evolve throughout the developing world. The book also considers the equity, health, and environmental effects of urban energy transitions. Critical to our assessment is the role public policy plays in the welfare of residential energy consumers and in the evolution of urban energy markets.

Why Examine Urban Energy Transitions?

The question can be raised: "Why undertake a study of urban energy transitions in the developing world?" There are several justifications. Developing country governments often invest heavily in infrastructure in the modern fuel sector, and they also frequently implement demand-side policies that affect energy pricing and consumer access. Extensive government involvement in developing country energy markets inherently raises policy issues.

There are also equity issues associated with urban energy transitions. Lower-income residents rely to a larger extent on traditional fuels than do higher-income consumers, and lower-income residents are disproportionately burdened by the costs (both pecuniary and nonmonetary) of residential energy utilization. This segment of the population is most vulnerable to policy changes instituted in energy markets. For these reasons, it is important to assess the distributional burdens associated with urban energy transitions and to consider the intended effects and possible unintended side effects of policies implemented in residential energy markets (Estache et al. 2001, World Bank 1996b).

Finally, negative externalities are associated with urban energy markets. The harvest and utilization of biomass-based fuels can accelerate deforestation and its associated environmental side effects (Wallmo and Jacobson 1998). In addition, the asset value of biomass stocks is not always reflected in harvest decisions when property rights are not well defined, leading to excess extraction (Hartwick 1992). The health consequences of particulates and other emissions from the combustion of traditional fuels is also receiving increasing study in the literature (Kammen 2001; Smith 2002; Smith and Mehta 2003; Smith 1993; Smith et al. 2000). Of particular concern is the exposure risk for women and children—the segment of the population that spends the largest amount of time around cooking fires.

Because of these policy concerns, an extensive body of research has developed on the subject of urban energy transitions. It is fair to say, however, that consensus has not been reached in the field on many important issues. For example, different conclusions have been reached about the factors that drive interfuel substitution and the income switch points at which consumers transition to higher grade fuels. There is also debate in the literature about the impact of energy prices on low-income consumers and the efficacy of different policies for encouraging interfuel substitution. Part of the reason behind this lack of consensus is the fact that conclusions in the literature have generally arisen from extrapolations of results from individual studies of single cities or a few cities (e.g., Adegbulugbe and Akinbami 1995; Alam et al. 1985; Barnes 1990; Bowonder et al. 1987a; Chauvin 1981; de Martino et al. 1991; Dewees 1995; Foley 1987; Hosier 1993; Hymen 1985; Leach 1986; Reddy and Reddy 1984; Sathaye and Meyers 1985; Sathaye and Tyler 1991; Sharma and

Ramesh 1986; Tibesar and White 1990).[1] This body of work on urban energy transitions has established a substantial knowledge base about local conditions in study regions, but it has yet to develop a consistent understanding of urban energy transitions as they occur throughout the developing world.

Overview of Research Methods

This book provides a coherent view of urban energy transitions and the associated policy options for intervening in urban energy markets. Our study is based on an integrative analysis of the results of a research program conducted under the auspices of the Energy Sector Management Assistance Program (ESMAP) of the World Bank over the period 1984–2002.[2] This program financed household energy studies based on interviews with more than 25,000 households in 45 cities as a part of surveys in 12 countries spanning 3 continents. The compilation, standardization, and analysis of this information produced a rich dataset for comparative analysis.

ESMAP originally was formed to deal with the after-effects of the rise in petroleum prices in the early 1980s. The original program financed national energy assessments in many developing countries. As the assessment work progressed, analysis revealed that household and biomass energy comprised the predominant fuel in many developing countries, even though these types of energy were ignored by the majority of the published research. This omission was the case for both rural and urban areas. Due to the varying household energy policies in urban areas, the program began to finance household energy strategies for developing countries, to complement the more general work at the country level. As a consequence, ESMAP began to focus on issues involving the efficient use of biomass fuels, local forest resource management around cities, the rural–urban biomass market chain, urban interfuel substitution, and other policy issues as they related to urban energy. This book is a partial summary of the valuable insights gained during that fertile period of energy research for developing counties.

A significant part of our study estimated periurban forest stocks around a subsample of 34 cities. In many countries, an adequate inventory of standing biomass has never been completed. Consequently, we developed a general methodology for estimating periurban forest stocks from secondary source information, and we used the methodology to generate a standardized dataset. Computerized mapping software was used to digitize site-specific vegetation patterns and to generate biomass density maps. This resource inventory allowed us to connect the conditions in local energy markets to periurban resource regimes and to compare information across our sample on the effect of urbanization on the periurban environment.

We also generated information on selected variables that can affect the availability of biomass fuels, stand degradation, and deforestation around urban areas. These variables included road infrastructure, topography, and precipitation.

In addition to the comparative analysis that forms the largest part of this book, we include a detailed analysis of an ESMAP-sponsored household energy survey of a particular city—Hyderabad India. Results are compared with those of a study of the same city conducted during the period 1981–1982 (Alam et al. 1985b). This longitudinal perspective allows the documentation of the evolution of urban energy choice, fuel consumption, and periurban forest resources over a 10-year period, providing a complementary perspective to the comparative study in other parts of the book.

In summary, the broadly based, synthetic assessment in this book offers a clearer and more comprehensive picture of urban market evolution than previously existed in the literature. It is our hope that this perspective will help guide future research on urban energy transitions and policy formulation in urban energy markets.

Cities and Regions in the Study

The cities included in our study are widely distributed in Africa, Asia, Latin America and the Caribbean, and the Middle East (Table 1-1). They vary significantly along a number of dimensions. Nearly half of the Asian cities in the study had populations greater than one million at the time the household energy surveys were conducted, including such large metropolitan areas as Manila, Bangkok, and Jakarta. On average, the African cities in the study were substantially smaller; only Harare and Lusaka had populations larger than one million. The sizes of other cities in the study tended to range between the largest Asian and smaller African cities.

A number of the cities in the sample are in the early stages of energy transition, marked by relatively high per capita biomass fuel consumption. The consumption shares for biomass fuels (fuelwood and charcoal) are greater than 90 percent for all the cities in Burkina Faso, for example, while at the other extreme, modern fuels provide more than 75 percent of the energy needs in several larger Asian and Latin American cities (e.g., Bangkok, Thailand, and La Paz and Oruro, Bolivia).

Agroclimate and biogeographical conditions also vary widely across the sample. For the most part, the Southeast Asian cities lie in moist, humid tropical regions, with mean annual precipitation levels ranging upward to close to 2,000 mm. The virgin closed-canopy forests of Southeast Asia are among the densest in the world, with biomass ranging up to 345 m^3/ha. In contrast, cities in two of the African countries in the study, Botswana and Mauritania, are

TABLE 1-1. Country and City Surveys Analyzed in Urban Household Energy Transition

Region	Country	City
Africa	Botswana[a]	Francistown, Gabarone, Selebi-Phikew
	Burkina Faso[b]	Ouagadougou, Bobodiougou, Koudougou, Ouahigou
	Cape Verde[c]	Mindelo, Praia
	Mauritania[c]	Altar, Kaedi, Kiffa, Nouadhibou, Nouakchoot
	Zambia[b]	Lusaka, Kitwe, Luanshaya, Livingstone
	Zimbabwe[c]	Bulawayo, Harare, Masvingo, Mutare
Asia	India[c]	Hyderabad
	Indonesia[c]	Bandung, Jakarta, Semarang, Surabaya, Yogyakart
	Philippines[c]	Bocolod, Cagayan de Oro, Cebu City, Daveo, Manila
	Thailand[c]	Ayuthaya, Chiang Mai, Bangkok
Latin America and the Caribbean	Bolivia[c]	La Paz, Oruro, Quillacollo, Tarija, Trinidad
	Haiti[c]	Port au Prince
Middle East	Yemen[c]	Sanna, Taiz, Hodeida

[a] Standing biomass data are available.

[b] Urban energy consumption data are available.

[c] Both standing biomass and urban energy consumption data are available.

located in semi-arid and arid regions (the Kalahari desert lies in southern and western Botswana); precipitation levels in Mauritania, the driest area in the study, range from only 35 to 500 mm per year. Open montane woodland and dry bush savannah in Zimbabwe have biomass densities ranging in the neighborhood of 62 m³/ha and 33 m³/ha, respectively. Precipitation levels and natural biomass densities in Haiti and Latin America fall within the levels encountered in Africa and Asia.

Topography also varies significantly among the study regions. The terrain in Africa is relatively flat: few areas in the study have average slopes greater than 8 percent. In contrast, the regions around the Southeast Asian cities in the study are far more mountainous, with substantial tracts having slopes greater than 8 percent. Topography is also varied in Bolivia: two of the Bolivian cities, La Paz and Oruro, are located in the Andes.

The sample of cities studied encompasses a sufficiently wide range of variation in populations, energy transition stages, biogeographies, and policies to enable generalization about the common factors underlying the evolution of urban fuel markets.

Topics of Study

The transition from traditional to modern fuels is important for urban people because of its potential to improve the quality of energy service, to lower indoor air pollution, and to stem deforestation pressures in periurban environments. This book investigates the socioeconomic and policy factors that shape urban energy transitions and the associated effects on equity, human health, and the environment.

Urban Energy Transitions

It is important to understand the factors influencing the relationship between urbanization, fuel choice, and energy consumption in developing countries, because this understanding provides the informational basis for policy interventions in urban fuel markets. Urban energy transitions are often discussed in the literature as smoothly sequenced evolution from firewood, to charcoal and kerosene, and ultimately to LPG and electricity consumption. In contrast, our study finds that the urban energy transitions are quite varied—in terms of the timing of the transition period and in the transition fuels consumed. It turns out, in fact, that the intermediate stage of the transition follows one of several distinct pathways. It is also the case that the income threshold at which people switch to modern fuels differs widely in different countries, depending on urban-specific household characteristics, resource conditions, and policy regimes. Consumers respond to energy price signals and constraints in urban fuel markets, and governments' roles in influencing energy prices and access have thereby played a crucial role in urban market evolutions. Biomass supply around cities is another variable that shapes the conditions found in urban energy markets, and it diversifies the expression of the urban energy transition in different cities and countries.

Equity

The poor in urban areas of developing countries face special problems in meeting their basic energy needs. Many of the urban poor are recent migrants from the countryside and continue to rely on traditional fuels they previously collected themselves. Because the opportunity cost of their time is generally low, the urban poor face a new financial burden when they begin paying the urban market price of traditional fuels. A number of inequities in urban energy markets also penalize low-income consumers. In some countries, poorer households pay substantial portions of their incomes for traditional energy, because they have limited access to such alternative fuels as kerosene, LPG, and electricity. Modern fuels, and/or the appliances needed to use them, may not be available in the marketplace due to inconsistent or restrictive gov-

:nment policies or the relative remoteness of urban locations. If substitute :els are not available, poorly functioning markets can increase the price con- imers pay beyond the economic costs of producing and delivering the fuels.

Our study finds that poorer people are paying higher prices for useable energy than more well-off consumers because of the inefficiency of tradi- tional-fuel-using cooking stoves and kerosene lamps. Due to information imperfections in the market, the relative inefficiencies of traditional-fuel- using appliances are not fully discounted in market prices. Poor people also spend a larger share of their cash income on energy than do wealthier urban consumers. As the principal consumers of traditional fuels, the poor also bear a disproportionate share of the health and inconvenience costs associated with residential energy consumption. Taken together, these findings suggest that the poor are relatively burdened by the pattern of residential energy uti- lization in developing countries. This inequity poses a fundamental challenge for policymakers.

Environmental and Health Effects

The impact of urban energy transitions on human health and the environ- ment is an important issue. To borrow from the recent Kuznets curve litera- ture (Panayotou 2003), "technology" and "scale effects" are among the critical factors that should affect the health and environmental result of urban eco- nomic expansion. A transition to modern fuels has the potential to reduce health and environmental effects through the first of these channels. Specifically, fuel switching to LPG and electricity should reduce pressure on periurban biomass stocks, while reducing indoor air pollution associated with emissions from burning traditional fuels. On the other hand, the expan- sion of urban areas has the possibility to increase the scale of aggregate demand for biomass fuels and the total exposure risks to indoor air pollution, even if a significant portion of the population switches to more modern fuels. In that case, urban development would exacerbate environmental and health problems for a longer period of time, until a more or less total market pene- tration of modern fuels. The transition time from biomass to modern fuels is thus an important variable ultimately affecting health and environmental effects. The consumption mix of different fuel types during the transition period will also affect the severity of the health and environmental effects experienced by urban residents during the urban market transition.

Our study finds that periurban deforestation is associated with urban growth in conjunction with the expansion of transportation infrastructure, combined with interactions with topographical and climatologic factors. Significantly, the per capita consumption of biomass fuels persists at a rela- tively high level until the advanced stages of the energy transition, and the aggregate consumption of biomass fuels does not necessarily decline with income growth. With total biomass energy consumption continuing at a high level as cities develop, the demand pressures on surrounding forested land will continue even after cities have reached the later stages of the modern fuel transition. And health effects from indoor air pollution will continue—particularly for low-income people.

In sum, none of the cities in our study have reached the developmental stage encountered in developed countries in which urban energy consumption is disassociated from the effects on periurban forests and from the health effects of traditional fuel burning. The tipping point for this particular environmental Kuznets curve is beyond the highest income level in our sample. However, we do find that policies can influence the timing and duration of the energy transition, as well as the compositional mix of fuels consumed during the transition period. Thus, policymakers have considerable scope for affecting the health and environmental impact of energy market evolution.

Definitions and Terminology

A number of terminological distinctions are employed throughout this book that deserve clarification at the outset. The term "modern fuel" is used to denote either LPG or electricity. The terms "woodfuel" and "fuelwood" are used interchangeably to refer to unprocessed firewood, which is distinguished from charcoal—a unique fuel on a number of grounds. Although charcoal is often used as a basic cooking fuel at the earlier stages of the energy transition, it can also serve as a transitional fuel in the intermediate consumption stage between woodfuels and modern fuels. In fact, charcoal is sometimes used in the latest stages of the energy transition as a specialty fuel for grilling. In that context, charcoal plays the role of a substitute for modern fuels. In view of the ambiguous definitional status of charcoal, the approach in the book is to semantically differentiate between charcoal and woodfuel when the distinction is relevant, while lumping these fuels together under the rubric "traditional fuels," "biomass-based fuels," or "biomass fuels," when referring specifically to traditional patterns of energy utilization at the earliest stage of the energy transition. Coal and kerosene, in contrast, are denoted as "transitional fuels" throughout the book, because their role as intermediaries in the evolution from traditional to modern fuel consumption is fairly evident.

A conceptual distinction is often made in the book between the choice to use a particular fuel, on the one hand, and its consumption level, on the other. The distinction is particularly relevant for the modern fuels, the use of which requires an initial capital outlay (in the form of the purchase of bottles, modern stoves, and/or the payment of hookup fees). In this case, the initial choice to use the fuel and the subsequent decision to consume a particular amount constitute distinct decisions in a two-stage consumer choice process.

However, for the traditional fuels, which involve lesser degrees of initial capital outlays, the fuel choice and consumption decisions are conceptually more conflated. In the traditional market setting in which households use woodfuel or charcoal, the distinction between choice and consumption probably reaches close to the absolute convergence implied in the standard microeconomics theory of price-rationed commodity demand.

Throughout the study, the choice to use a fuel is measured as the percentage of households that consume a given fuel type, while the level of fuel consumption is measured in standardized quantity units of kilograms of oil equivalent (kgoe) per month. We explicitly note cases in which this standardized unit is further adjusted to reflect the efficiency of fuel-using appliances, thereby giving the energetic measure in units of useful delivered energy. This measure is the actual heat used in the process of heating and cooking food.

The choice to use a fuel can be further distinguished from "fuel access." Modern fuel markets in many urban areas in the developing world are access rationed—for example, only certain customer classes may have access to electrical hookups. Access to a fuel is a necessary, but not sufficient, condition for choosing to consume the fuel.

The distinction between access, choice, and fuel consumption are sometimes blurred in the book when not particularly relevant—for example, we have referred generically in this chapter to the "urban energy transition" without making the terminology distinction until now. In that case, the reader can assume that we are implicitly referring to all measures, that is, an energy transition that features greater access to modern fuels, greater market share of modern fuels (in terms of the fraction of consumers choosing to consume them), and an increasing level of modern fuel consumption. The distinction is sharpened when it becomes particularly relevant; specifically, in the analysis of household consumption behavior in Chapter 3. The term "fuel-switching" definitionally refers to the choice to use a new fuel, but the degree of fuel switching is gauged in the study both as the fraction of the customer base choosing to consume particular fuels and to the associated consumption levels.

Finally, the terms "price" and "market price" in the book refer to the market prices of the different fuels normalized to reflect different purchasing power parities and energy content. Specifically, the standard price unit is 1988 dollars per kgoe. In some cases, we find it useful to compute relative prices. The local price of kerosene is used as the reference fuel in these cases because it is available in all countries and is a transition fuel between the traditional biomass fuels and the modern fuels. Income and expenditure measures are also in 1988 dollars and reflect the standard methodology for normalization used at the World Bank.

Methodology Caveat

The data in our study are derived from a "bottoms-up" compilation and standardization of primary source household energy survey data collected over a number of years, as well as from secondary source information yielding estimates of periurban biomass inventory. As we have suggested, this database represents a substantially more comprehensive informational foundation than found in the existing literature. Notwithstanding, our dataset has lower information content than a complete panel dataset would have. But panel data on the variables of interest are not available, because governments do not maintain standardized statistical databases on household energy consumption, indoor air pollution, or periurban resource characteristics.

Due to the type of information we have, our study perforce relies on less formal methods than the standard statistical analysis of panel data. We compared averages and evaluated partial correlations, and we drew qualitative inferences and deductions based on our own professional experience. Because the conclusions of this book are not formally tested, they must be regarded as best available estimates given the state of information and the authors' judgment. Obviously under these circumstances, our study does not provide the last word on the big subject of urban energy transitions. The contribution of this book should itself be regarded as a transitional stage in the state of research on urban market evolution.

Attention today is just turning to the inclusion of energy issues in the large multisector surveys now taking place in developing countries. These surveys include the World Bank's living standard measurement surveys and the demographic and health surveys funded by the U.S. Agency for International Development. In the coming years, these new sources of information may be able to provide a better foundation for a more formal statistical analysis of the patterns and trends that have emerged from the research conducted in this study.

Conclusions

Our study shows that few societies have traversed a regular and consistent path from traditional fuel consumption to the use of electricity and other modern fuels. A variety of transitional consumption pathways are followed, and markets evolve more or less rapidly and directly towards modern fuels. The complexity and diversity of the transition process, combined with the piecemeal, case approach to its study, has given rise to many viewpoints in the literature with respect to the important determinants and permutations. In our view, the incompleteness of this picture has complicated the task of policy formulation in urban energy markets. The goal of this book is to broaden and deepen the foundation for understanding urban energy markets, in the hopes of helping policymakers to improve strategies that assist consumers to use energy more efficiently and equitably and that reduce the health and environmental side effects of urban energy transitions.

The following chapters of this book examine distinct features of the urban energy transition. Chapter 2 provides an overview of urban energy transitions. The chapter develops a conceptual model for urban market evolution, and it examines empirical evidence for the relationship between the size of cities and energy market development. Evidence is also considered for the relationship between resource characteristics around cities and fuel consumption profiles, as well as the effects of policies on urban energy markets in different cities and countries. A taxonomy of different kinds of energy transitions is developed.

Household energy consumption behavior in urban energy markets is examined in detail in Chapter 3. Key attributes are shown to be household characteristics, including size and family income, as well as energy price signals and access restrictions. In combination, the study shows that household characteristics, incentives, and constraints—in the form of price signals and access restrictions—determine residential fuel choice and consumption in urban markets. This information is important for understanding the way policy can affect consumer behavior and urban energy transitions.

The equity implications of urban energy policies and their implications for market evolution are detailed in Chapter 4. The analysis is based on a dissagregation of fuel choices, consumption patterns, and energy expenditures by 10 income classes. The price of energy for lower-income residents is assessed, as well as the responsiveness of different income classes to market signals and policies. The policy implications of inequitable residential fuel choice and consumption patterns are considered.

The environmental and human health implications of urban energy transitions are examined in Chapter 5. This includes a unique analysis of the impact of urbanization on land-use patterns surrounding urban centers.[3] The aggregate level of traditional fuel consumption is computed for each of the cities in the sample. The implications of aggregate consumption patterns on periruban environments and human health are assessed.

There is a detailed analysis of a specific household energy survey conducted in Hyderabad India, and the results are presented in Chapter 6. This case study assesses the topics studied in Chapters 2–5 within the context of a single urban area and, by comparison to an earlier study of the same city, it provides a longitudinal picture of urban energy consumption patterns and resource conditions over a 10-year period.

Finally, Chapter 7 provides a summary of the results and a discussion of policy implications. The chapter identifies particular policy issues and solutions tailored to each stage of the urban energy transition. Although there are many issues to consider in formulating policy in urban energy markets, a general suggestion that emerges from the study is to use flexible policy approaches that maximize consumer choice. This strategy appears to be the best means to encourage economic efficiency and equity in urban energy markets and to minimize the health and environmental side effects of urban energy transitions.

2

The Urban Energy Transition

Markets for goods and services are relatively underdeveloped in the earliest stages of a city's evolution, and institutional arrangements are relatively informal. As cities expand and modernize, the scale of economic activity increases, and urban institutions become more complex. Paralleling this transition is an evolutionary pattern of consumer fuel choice and home energy consumption. As cities grow beyond their earliest developmental stage and ultimately modernize, households generally shift away from the use of traditional biomass fuels and end up consuming LPG and electricity.

The transition from traditional to modern fuels has often been conceptualized in the literature as a relatively straightforward, three-stage process. In the first stage, woodfuel is the predominant energy source. The second stage is marked by local deforestation, decreasing wood availability, and the emergence of markets for such "transition" fuels as charcoal and kerosene (see Boberg 1993; Hosier and Kipondya 1993; Malawi 1984; Milukas 1986). Finally, the third stage is characterized by developed markets, rising incomes, and large-scale fuel switching to LPG and electricity (Leach 1993).

However, our study shows that this conventional view is an oversimplification. The analysis starts in this chapter with a conceptual overview of the urban energy transition process. It presents some data trends that shed light on the general features—as well as particular variations—in energy choice and consumption behavior that can emerge as cities develop and modernize. We then use factor analysis to classify cities in the study according to their energy market profiles. This classification shows that the cities in the study have not undergone a smoothly sequenced transition from fuelwood, to charcoal and kerosene, and then to LPG and electricity consumption. Particularly in the middle stages of the transition, more diverse fuel choices and consumption patterns are evident. These variations can be identified as distinctive substages in the transition process. The possibility of distinctive transitional patterns has policy implications with respect to consumer welfare, equity, and health and environmental effects of urban energy transitions.

Conceptual Model of the Urban Energy Transition

Urbanization is not simply an increase in population density but also a process of fundamental transformation in the organization of human behavior. Part of the influence of urbanization—both in the sense of migration to cities and growth of the cities themselves—is the loss of rural habits and traditional behavior patterns, and the acquisition of new domains of information and infrastructure.

The relationship between urbanization, fuel choice, and household energy consumption is a dynamic process involving a complex set of feedbacks. This complexity gives rise to the possibility of a variety of transitional pathways in urbanizing energy markets (Barnes 1990; Barnes and Qian 1992; Bowonder et al. 1987b; Dewees 1989; Hosier and Kipondya 1993; Millington et al. 1990; Taun and Lefevre 1996). In the prototypical case, wood is extensively available around cities in their earliest stages. Woodfuel harvest and production is a labor-intensive process, and the opportunity cost of labor time is generally low. Consequently, traditional fuels can be supplied relatively economically. Woodfuel stock is also readily available at this stage as a side effect of agricultural land conversion (Barnes and Qian 1992). In contrast, modern fuels will usually be hard to obtain and/or costly, because the urban area is not large enough or wealthy enough or close enough to the main distribution lines to attract commercial energy suppliers. The price of woodfuel at this stage may range from very low to relatively high—the latter due to seasonal scarcity or lack of access to modern fuel alternatives, among other factors. Regardless, the relative price of woodfuel is lower than the backstop price of modern fuels.[4] Consequently, urban residents typically consume woodfuel to the exclusion of other fuels at the beginning stages of a city's development.

Even so, the consumption of biomass fuels will vary across cities at this developmental stage, depending on the pattern of land use surrounding the city, the level of rainfall, the rate of stock regrowth, and other environmental factors. Moreover, policy interventions may also influence the relative economies of traditional fuels versus modern alternatives, for example, the presence or absence of rural electrification programs.

The incentive to consume biomass fuels will be moderated by a number of feedback relationships as urban areas expand. Over time, land clearing for agriculture and forest cutting for biomass fuels (and other forest-derived products) will decrease natural biomass stocks, reducing the volume and accessibility of natural biomass within the urban perimeter (Barnes 1990).

Transportation infrastructure and rural population densities in the vicinity of growing cities will also expand, increasing pressure on the periurban resource base. This evolution may be moderated by reforestation projects, market-driven reforestation activities if tenure regimes are secure, and natural stock regrowth if climates are favorable (Dewees 1989). In general, however, biomass resources in the vicinity of cities will be diminishing as urbanization proceeds, increasing the harvest and transport costs of woodfuel in particular.

The modern fuel sector will also grow as urban areas expand, due to economies of scale in infrastructure development and fuel distribution, which makes bottled LPG distribution and the expansion of electricity grids more economical (Leach 1993). Urban growth also increases tax revenues and the scale and sophistication of financial institutions. Among other consequences, "financial deepening" will augment the financial resources available for large capital projects, such as power generation (Darrat 1999).

As urban areas develop, biomass supplies are depleted in a larger radius around the city, and the relative price of woodfuel may increase due to increased scarcity, as well as to rising transit costs. Modern fuels will become more available and affordable through well-established distribution networks. Rising incomes and the relative availability, affordability, and superiority of modern fuels will encourage a consumption shift away from traditional fuels. Indeed, competitive price pressures from a modernizing fuel sector may ultimately lower the market price of urban energy, including resource-constrained traditional fuels, if fuel markets are freely functioning (e.g., if they do not face government-imposed access or quantity constraints).

How rising incomes are distributed is one of the several factors that can inject variation into the pattern of urban energy evolution. Rising but inequitably distributed income can lead the wealthiest class in society to rapidly transition to modern fuels while the lower classes remain dependent on traditional fuels. Rising but more equitably distributed income is likely to yield a more broadly based and less abrupt transition. Under these conditions, a substantial middle class is likely to serve as a demand source for a combination of fuelwood, charcoal, and kerosene for a more prolonged period of time. Depending on fuel availability and household income, some of these households may also use LPG and electricity. On the other hand, stagnating incomes will slow the transition, inducing urban residents to continue to use fuelwood or charcoal, and therefore prolonging demands on periurban resources.

The rate of urbanization is another factor that can affect the evolutionary development of urban energy markets (Hosier and Dowd 1988). Rapid urbanization initially increases the total urban demand for biomass fuels, because rural people who migrate to cities tend to maintain their traditional habits, purchasing biomass cooking fuels from vendors who harvest fuelwood from surrounding woodlands (Allen and Barnes 1985). However, by causing rapid deforestation around cities in the early stages, this pattern may actually accelerate the transition time between the period of traditional fuel consumption and the use of modern fuels. Rapid urbanization may feed a compensatory demand surge for commercial fuels including kerosene, LPG, and electricity at the later stages. This surge sometimes outruns the supply capacities of the local LPG or power companies when these fuels are quantity rationed, resulting in fuel shortages.

For cities with inequitable income distribution and rapidly growing populations, the transition to modern fuels may be prolonged for the lower income classes. On the one hand, even in remote cities at the earliest stages of the transition, the highest income families are likely to have some access to modern fuels, such as trucked-in petroleum products, and the equipment, such as generators and appliances, to use these products. This class may remain largely uninfluenced by the availability or price of local biomass supplies. However, poorer households will continue to depend on the biomass resources surrounding the city.

Government policy is another factor influencing the evolution of urban energy markets. The implementation of pricing policies, quantity rationing, or import controls can alter the pace and expression of the urban energy transition. Kerosene subsidies can encourage more rapid fuel-switching away from wood and/or induce consumers to continue kerosene consumption for an extended time period, while kerosene taxes can have the dichotomous effect of delaying the energy transition for low-income consumers but accelerating the switch time at which higher-income people choose LPG or electricity. Government policies also influence the market penetration of modern fuels in larger cities through access and/or quantity constraints (see Chapters 4 and 6).

Some Empirical Findings

We now consider some results that suggest the way in which the cities in our study have reflected the urban energy transition. As expected, the data show that the percentage of consumers who use traditional fuels declines, and the percentage choosing modern fuels rises—moving from the smallest town to the largest cities in the study (see Table 2-1). Per capita consumption trends parallel the fuel choice trends. However, the data suggest that the trends are not linear. Traditional fuel utilization does not decrease in cities where populations have not yet reached one million. Before this point, many location-specific factors, including environmental conditions and government policy, diversify fuel choices and consumption patterns. Beyond this population level, urban residents in cities fuel switch from biomass to modern fuels (and charcoal to some extent). In the million-plus cities in the study—including Manila, La Paz, Bangkok, and four cities in Indonesia—households choose to use charcoal infrequently, and they use almost no fuelwood for cooking.

Biomass resources around cities are an important source of variation in the sample. "Biomass resources" are proxied by an index that reflects the volume of woodfuel stock as well as its distribution density and proximity to cities. The index is developed in a straightforward way from the distance variables and the standing biomass stock data analyzed in Chapter 5.[5]

The data show that both the percentage of consumers who choose woodfuel for cooking and the absolute level of woodfuel consumption continuously decline with diminishing biomass resources (see Figure 2.1). In periurban areas in which wood stocks are abundant, per capita consumption ranges between 5 and 6 kgoe per month, dropping to less than 2 kgoe per month in areas of relative biomass scarcity.

Charcoal is also widely used in urban areas with plentiful wood resources, although consumption is slightly lower than for woodfuel use. However, the percentage of people using charcoal diminishes at a relatively low rate as wood resources become more scarce. The per capita consumption of charcoal actually increases in regions with adequate resources and stays close to constant as resources become scarce (see Figure 2.1). As a result, charcoal consumption dominates fuelwood consumption in urban areas with only adequate or scarce biomass resources. The relative difference in these fuel consumption trends can be explained by two factors. First, charcoal, as the lighter-weight fuel, can be trucked more economically for longer distances

TABLE 2-1. City Size and Energy Use in 45 Cities, 1988

City type	Population ('000)	Monthly income (US$/capita)	Fuelwood	Charcoal	Kerosene	LPG	Electricity
			Fuel				
			Energy consumption (kgoe/capita/month)				
Town	33.89	38.19	3.82	3.33	0.21	1.70	1.41
Small city	102.54	41.38	2.19	2.15	0.62	2.12	1.59
Middle city	526.98	35.74	3.41	3.08	1.40	0.60	1.27
Large city	3,718.13	55.82	0.24	1.24	3.35	1.68	2.82
			Energy choice (percentage)				
Town	33.89	38.19	52.50	40.00	33.60	46.50	64.10
Small city	102.54	41.38	25.10	36.10	37.20	60.40	78.40
Middle city	526.98	35.74	47.90	53.30	64.50	23.00	69.50
Large city	3,718.13	55.82	4.30	28.00	61.30	37.30	95.40

Sources: World Bank 1988, 1989, 1990a, 1990b, 1990c, 1990d, 1991a, 1991b, 1992, 1993, 1996a, 1999 (hereinafter "ESMAP Household Energy Surveys").

than can woodfuel. Hence, charcoal imports from more distant regions can economically replace locally produced sources as local biomass resources are depleted. Secondly, charcoal is a superior fuel. In fact, the demand for charcoal for specialty cooking continues even at the latter stages of the energy transition.

As mentioned, government policies can play an important role in urban energy transitions. The data show that both subsidies and taxes are correlated with the extent to which fuels are used in the cities in our study. For example, coal subsidies in China are associated with high coal consumption levels (left panel of Figure 2.2). Highly subsidized coal in China is in fact the lowest-priced fuel in all of the developing countries in the study. However, coal consumption there is not just due to the subsidy. It also reflects the fact that the government also distributes coal for heating. Thus, coal is a ubiquitous and familiar fuel source to Chinese consumers. In Indonesia, government policy subsidizes kerosene to assist poor households and to prevent deforestation, leading most people to use kerosene for cooking. This high kerosene consumption pattern is unique among the 12 countries in the study. Fuel subsidies also affect the prices of alternative fuels. As the data in Table 2-2 indicate, the prices of fuel alternatives in countries that subsidize a major fuel are lower than in the other countries in the study.

Taxation is correlated with a decreased usage of the taxed fuels and an increased usage of fuel substitutes. Taxes on imported petroleum in Haiti, Mauritania, and Burkina Faso have pushed consumers in these countries to utilize fuelwood or charcoal (right panel of Figure 2.2). In fact, Burkina Faso's LPG price is the highest in the study, at $1.20 per kgoe—about four times the international price. Urban consumers in the country also have the highest per capita consumption of wood of any in the study. Such tax regimes, as well as foreign exchange constraints on imported petroleum products, contribute to this heavy reliance on traditional fuels in these countries.

Cities in countries with more laisser faire market-based pricing regimes evince a more mixed pattern of fuel consumption than do cities in countries in which energy markets are skewed by government intervention (center panel of Figure 2.2). A notable case is Cape Verde, a small African island nation in which urban households use relatively high proportions of electricity, LPG, and kerosene. The general pattern of mixed fuel use in undistorted market economies holds true across countries, demonstrated by the energy consumption figures for the middle-income groups in countries without taxes or fuel subsidies (Table 2-2).

In addition to incentive-based policies such as taxes and subsidies that influence consumer behavior through effects on energy prices, governments also directly influence fuel choice and consumption through access rationing. Access rationing of electricity is quite common in a number of countries in the study. In Mauritania and Burkina Faso, for example, access to electricity

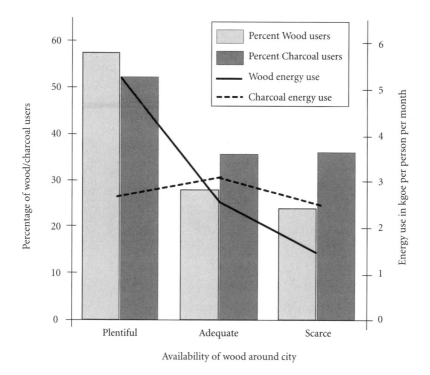

FIGURE 2-1. Impact of Availability of Wood on Use of Wood and Charcoal

is restricted to higher-income urban residents. Access rationing has an immediate effect on the excluded customer class, thereby raising equity issues, and
it also affects the functioning of energy markets. The effects of access
rationing on energy choice and equity is further discussed in Chapters 3, 4,
and 6.

Urban Energy Transition Typology

The description of the conceptual model and the overview of the data trends
suggest the possibility of a variegated transition process as a function of differences in biomass resources and government policies. We now ask if it is
possible to be specific about the energy transition pathways followed by the
cities in our sample. We use factor analysis in an attempt to classify cities
according to some common elements. The variables included in the factor
analysis are fuel choice (percentage of households in the city who consume a
particular fuel), fuel consumption share (percentage of the city's energy consumption total accounted for by a particular fuel), average energy prices faced

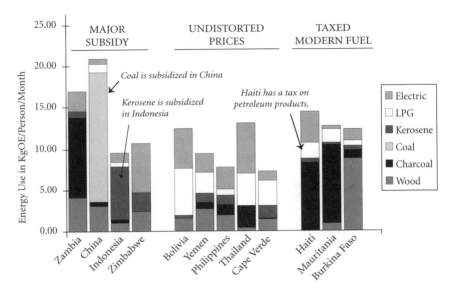

FIGURE 2-2. Energy Use and Government Policy

by the consumers in the city, wood resources around the city, and average per capita income of residents in the city. The stages in the urban energy transition are derived from the main factors generated by the analysis and the factor scores of the cities. The percentage of households in the city that use the different types of fuels—the fuel choice measure—is indicated in Table 2-3, while Figure 2.2 indicates fuel consumption shares. At this aggregate city level, the typology by fuel choice and fuel consumption share give consistent rankings. The data illustrate the variations in the transition pattern and, in particular, the distinct groupings into which the second stage of the transition can be subdivided.

Stage 1: Utilization of Biomass Fuels

The first group of cities in the study can be characterized by their extensive use of unprocessed wood as a cooking fuel (Table 2-3 and Figure 2.3). These cities are relatively underdeveloped economically, and woodfuel production and distribution costs are low due to some combination of low wage rates and substantial biomass resources around the cities. The remoteness of locations in many cases also reduces the cost effectiveness of modern fuel imports. The relative price of traditional fuels varies substantially among these cities, depending on local resource conditions, policies, and availability of fuel imports. For example, the city of Xiushui, China, is located in a mountain valley with extensive wood resources from nearby forests, and the price of woodfuel is relatively low (World Bank 1996a). In contrast, the price of woodfuel

TABLE 2-2. Relationship between Fuel Policy, Energy Pricing, and Fuel Consumption in 12 Countries

Energy measure	Monthly income ($/person)	Fuel					
		Fuelwood	Charcoal	Coal	Kerosene	LPG	Electricity
All income groups							
		Useful energy price (US$/kgoe)					
Subsidy	21.80	0.97	0.77	0.11	0.69	0.40	0.56
Undistorted	77.98	2.42	2.92	NA	1.11	0.61	1.47
Taxed	41.85	2.13	1.83	NA	1.58	1.31	2.63
		Energy consumption (kgoe/capita/month)					
Subsidy	21.80	2.61	3.38	5.19	2.30	0.46	1.24
Undistorted	77.98	1.25	0.63	0.00	0.73	3.06	2.94
Taxed	41.85	3.13	5.78	0.00	0.43	0.83	1.43
Middle-income groups only (US$20 to $40 per month)							
Useful energy price (US$/kgoe)							
Subsidy	26.32	0.91	0.95	0.10	0.67	0.39	0.60
Undistorted	28.94	2.36	2.56	NA	1.01	0.54	1.24
Taxed	29.03	2.62	2.05	NA	1.66	1.45	2.95
		Energy consumption (kgoe/capita/month)					
Subsidy	26.32	1.34	2.45	6.01	2.73	0.84	1.26
Undistorted	28.94	1.78	0.94	0.00	0.83	2.11	1.43
Taxed	29.03	4.97	4.77	0.00	0.42	0.70	0.42

Note: NA = not applicable.

Source: ESMAP Household Energy Surveys.

is quite high in Burkina Faso relative to other countries in the study, but fuel alternatives are also expensive—due in part to a tax on petroleum-based fuel. In the case of Burkina Faso, consumers in Koudougo, Ouahigouya, Bobo Dioulasso, and Ouagadougou still use relatively high-priced traditional fuels because the price of the taxed modern fuels is even more expensive. In Livingstone, Zambia, people use both woodfuel and charcoal for cooking. Although unprocessed wood is still available around the city perimeter, resource pressures and regulated charcoal prices are inducing residents to switch to charcoal as a cooking fuel.

Stage 2: Utilization of Transition Fuels

The second stage of the energy transition typically follows one of three main substage trajectories, depending in part on government policies toward transition and modern fuels. During this stage, cities may adopt a pattern of high

TABLE 2-3. Classification of Cities by Stage in the Energy Transition (percent of households consuming fuel types)

City rank	Income$/capita/month	City size '000s	Fuelwood %	Charcoal %	Coal %	Kerosene %	LPG %	Electricity %
Stage 1								
1. Bobo Dioulasso, Burkina Faso	30.42	247	94.4	18.5	0	1.0	4.5	0
2. Xiushui, China	9.32	40	95.0	5.0	55.0	27.0	21.0	0
3. Livingstone, Zambia	18.93	81	50.6	71.3	0	3.4	0	16.7
4. Koudougou, Burkina Faso	22.40	55	94.2	18.8	0	0	1.4	0
5. Ouagadougou, Burkina Faso	38.71	473	90.6	13.7	0	1.0	11.3	0
6. Ouahigouya, Burkina Faso	29.74	41	100.0	2.2	0	0	0	0
7. Davao, Philippines	20.24	839	59.2	51.0	0	38.8	24.5	12.2
8. Cagayan, Philippines	27.45	312	86.1	25.0	0	44.4	19.4	11.1
Stage 2A								
9. Kitwe, Zambia	27.72	360	40.2	79.9	0	8.7	0	41.4
10. Luanshya, Zambia	16.89	149	28.4	92.1	0	2.6	0	34.7
11. Kiffa, Mauritania	20.21	20	57.5	89.7	0	0	3.4	0
12. Lusaka, Zambia	28.89	704	15.4	78.0	0	24.2	1.1	32.4
13. Kaedi, Mauritania	15.77	12	93.7	68.4	0	0	8.9	0
14. Port au Prince, Haiti	65.08	1000	4.4	93.3	0	4.1	28.0	7.2
15. Atar, Mauritania	33.57	35	30.9	52.7	0	0	27.3	0
16. Nouadhibou, Mauritania	42.57	60	1.3	61.8	0	0	38.2	0
17. Nouakchott, Mauritania	24.44	550	9.0	89.2	0	1.6	39.2	0
Stage 2B								
18. Surakarta, Indonesia	15.11	688	30.3	11.4	0	62.1	6.8	6.1
19. Yogyakarta, Indonesia	27.88	645	24.4	13.6	0	69.9	2.8	4.5
20. Surabaya, Indonesia	22.26	2226	2.5	0.3	0	92.7	3.8	8.9
21. Semarange, Indonesia	17.12	1068	10.8	0	0	80.0	13.8	6.2

22. Bandung, Indonesia	26.40	2308	2.8	0	0	93.5	7.4	4.6
23. Jianyang, China	16.08	50	32.0	1.0	97.0	5.0	1.0	0
24. Jakarta, Indonesia	27.15	7976	2.9	0	0	91.5	10.5	5.9
25. Changshu, China	26.30	120	1.0	0	100.0	1	65.0	0
26. Kezuo, China	15.19	32	14.3	0	92.9	0	66.3	0
27. Huantai, China	22.67	55	1.0	0	85.9	0	97.0	0
Stage 2C								
28. Bacolod, Philippines	37.12	360	42.4	55.9	0	23.5	20.6	0
29. Cebu City, Philippines	32.02	674	35.2	25.9	0	16.7	33.3	0
30. Hodeidah, Yemen	41.21	182	20.5	10.3	0	43.6	61.5	0
31. Harare, Zimbabwe	37.21	718	9.4	0	NA	46.5	0	43.3
32. Trinidad, Bolivia	76.93	49	21.5	0	0	4.1	78.8	1.2
33. Bulawayo, Zimbabwe	62.62	451	9.7	0	NA	15.3	0	75.0
34. Manila, Philippines	67.86	8150	7.0	22.0	0	32.0	61.3	24.4
Stage 3								
35. Mindelo, Cape Verde	43.09	50	26.7	1.1	0	38.9	91.1	0
36. Tarija, Bolivia	70.21	74	9.9	0.2	0	1.0	90.4	1.0
37. Praia, Cape Verde	65.46	59	25.0	10.5	0	16.9	97.6	4.8
38. Quillacollo, Bolivia	59.11	36	21.8	0	0	0.4	92.4	1.1
39. Oruro, Bolivia	42.67	190	3.9	0	0	13.8	95.0	3.7
40. La Paz, Bolivia	78.37	1017	2.9	0	0	17.2	83.6	11.6
41. Tiaz, Yemen	48.47	161	30.0	6.7	0	10.0	96.7	0
42. Sanaa, Yemen	118.86	472	45.4	6.2	0	2.1	95.9	15.5
43. Ayutthaya, Thailand	84.45	40	9.4	45.5	0	0	82.1	76.4
44. Chiengmai, Thailand	102.09	150	17.5	64.0	0	0	69.2	73.0
45. Bangkok, Thailand	142.21	6000	1.6	24.5	0	0	89.6	81.6

Note: NA = not available in the survey.
Source: ESMAP Household Energy Surveys.

use of charcoal, high use of coal or kerosene, or a more diversified use of transition fuels.

Substage 2A: High Charcoal Use. The second stage often involves a substage in which households switch from fuelwood to charcoal. Urban dwellers in the cities in this substage face low relative charcoal prices for some combination of reasons: wood stocks are becoming scarce in the immediate vicinity of the urban area, raising the price of locally provided woodfuel; relatively low-cost charcoal can be imported from outside the area; or taxes on modern fuel substitutes or government-controlled charcoal prices help reduce the relative price of charcoal. In effect, people shift to charcoal as their predominant fuel source because alternatives are relatively expensive.

Our sample of countries illustrates the range of government policies that can enter at this substage. At the time of the Zambian household energy survey (World Bank 1990d), the Zambian government controlled the price of charcoal, making charcoal relatively affordable in relation to other fuels. In contrast, modern fuels were taxed in Mauritania and Haiti. These taxes provided a disincentive even for higher-income families in these countries, whose fuel demand was thereby displaced to charcoal. Charcoal prices were relatively high in these countries as a result of the policy-induced distortion, but they were still lower than the price of the modern fuel alternatives.

Substage 2B: High Coal or Kerosene Use. Another substage features cities in which kerosene or coal serves as the dominant transition fuel. In these cities, subsidies on kerosene or coal largely displace charcoal from the transition mix. As previously mentioned, coal is abundant in China, and policymakers subsidize it to improve the quality of life of people living in cities. Thus, woodfuel and charcoal have been virtually eliminated from household use (Table 2-3 and Figure 2.3). The city of Xiushui is the one exception to this rule, due to its unique location and available biomass resources. But as a general rule, people in China use coal and kerosene for cooking and heating.

An earlier study found that the government of Indonesia's subsidy of kerosene was large enough to appear to inhibit higher-middle-class families from switching to modern fuels. The price of kerosene varied around a very low average of 14 cents per kgoe. Without this policy, it seems likely that the higher-income families would be using LPG or electricity, and the poorer households would be using woodfuel or charcoal.

Substage 2C: Diversified Transition Fuel Use. The cities in a third substage are distinguishable by the use of a variety of fuels including LPG, electricity, charcoal, kerosene, and woodfuel. A major reason is that fuel markets are largely undistorted by government interventions, allowing local economic conditions to diversify fuel consumption patterns. The cities in this stage are

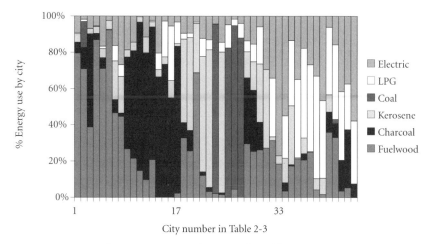

FIGURE 2-3. Fuel Consumption Share, Ordered by Energy Transition Stage

found in a variety of countries, including Bolivia, Zimbabwe, Yemen, and the Philippines. In the city of Trinidad, Bolivia, for example, people use both woodfuel and LPG for cooking. Trinidad has relatively higher incomes compared with other cities in the study, but it is also relatively small and remote from main population centers, with abundant local wood resources. Hence, woodfuel is used as well as LPG. In Bulawayo, Zimbabwe, there is little wood around the city, and people use kerosene, coal, or electricity for household use. Wealthier and some middle-class people already have made the transition to LPG or electricity.

Stage 3: Transition to LPG and Electricity

In the final stage of the energy transition, cities tend to be large, and a relatively high proportion of urban residents have relatively high incomes. Both electricity and LPG are the dominant cooking fuels in these urban areas. This evolutionary stage is well accepted in the literature.

The cities represented by this group in the study are in Thailand, Bolivia, and Cape Verde. Urban dwellers in Thailand, for example, have the highest per capita incomes in the study. Most people in Thai cities cook with LPG, but they also use electricity for many purposes. However, charcoal is still used for the preparation of traditional foods. Likewise, in some of the smaller, but high income, cities in Bolivia, people have access to relatively cheap fuelwood and still use fuelwood in conjunction with LPG.

The two countries of Cape Verde and Yemen are interesting because they have intermediate income levels, but they predominantly use LPG and electricity. In Cape Verde, a small island country, limited local wood resources provide about 20 percent of people's energy needs. The balance is accounted for by relatively affordable LPG imports. Yemeni cities have almost no wood resources. Hence, urban dwellers rely on bottled LPG or electricity.

Conclusions

The home energy needs of consumers are met by woodfuel during the earliest developmental stages of cities, but consumption ultimately shifts to LPG and electricity as cities grow and modernize. The evidence does not support the notion, however, that this transition from wood to modern fuels follows a regular pattern. Instead, it appears that energy transitions take varying amounts of time, and they feature different twists and turns along the way. The distribution of urban income, government polices, and the availability of wood resources are among the location-specific factors that appear to accelerate or delay urban energy transitions, as well as to determine the particular transition route that a city follows along the path to modern fuel utilization.

Although quite variable and complex, the evidence suggests that transitional fuel utilization pathways can be grouped into three distinct substages. In one substage variation, charcoal is the dominant transition fuel. The relative price of charcoal is lower than other transitional fuels, either due to local economic conditions or levied taxes on kerosene and LPG. In another variation, subsidies on coal or kerosene virtually eliminate the consumption of other fuels. Finally, there are a group of cities in which transition fuels are quite diversified, with wealthier households consuming LPG and electricity and poorer households consuming wood, charcoal, or kerosene. In these countries, relatively undistorted energy markets allow local economic conditions to differentiate a mix of cost-effective fuel options.

Government policies have played an important role in urban fuel utilization, especially in the middle stages of the energy transition. Policies instituted in energy markets may have varying effects, depending on the energy transition stage and market conditions in particular cities. These topics are discussed in later chapters.

3

Household Fuel Choice and Consumption

The actions of urban residents operating in a particular market environment determine their pattern of residential fuel choice and energy utilization. As we have seen, the kind of market urban consumers confront is heavily influenced by local resource conditions and the actions and policies of governments.

This chapter focuses in detail on the fuel choice and consumption decisions of urban energy consumers. The goal is to better understand why urban residents choose to consume the fuels they do, and how incentives influence their overall level of consumption. This information helps us better understand the way urban energy transitions unfold, and it offers insight into the effects of energy policies implemented in urban markets.

Several variables, particularly household size and household income, importantly influence the fuel choice and consumption decisions of urban residents. Price signals that consumers face in urban energy markets mediate the effects of the household characteristics that drive fuel demand (Bhatia 1985; Bowonder et al. 1988). Moreover, the availability of modern fuels in urban markets is often access-constrained. Because policy influences both consumer prices and fuel access, these two channels offer crucial entry points into the market through which government policy influences fuel utilization. All together, this group of influences—household attributes, market prices, and access constraints—determine the fuel choice and consumption decisions of urban residents.

TABLE 3-1. Fuel Choice and Total Energy Use by Size of Household

Average household size	Fuel choice (percentage of households choosing)					Total energy use in kgoe/capita
	Wood	Charcoal	Kerosene	LPG	Electricity	
Two	18	28	40	30	80	23.13
Three	13	16	22	53	93	17.65
Four	20	27	34	53	92	14.49
Five	21	36	47	48	87	10.44
Six	34	55	52	31	77	10.41
Seven	46	54	62	27	53	10.55
Eight	45	65	67	40	50	8.85
Nine +	73	37	66	24	42	8.33

Source: ESMAP Household Energy Surveys.

Effect of Household Characteristics on Fuel Use

Household characteristics, including family size and income, are important determinants of consumer demand. Recent migrants to urban areas often arrive as larger family units, and such families have a history of using traditional fuels. To some degree, this habit is retained in the new urban surroundings. This family-size-related consumption characteristic has not been fully assessed in the literature on urban energy transitions.

The income level of urban residents is another important household characteristic that influences the demand for urban fuel. Urban income is well recognized in the literature on urban energy transitions, although debate exists about the income point at which urban residents switch to modern fuels.

Household Size

Moving from a predominantly rural area to an urban area involves many changes for households. Large families are an economic asset in the countryside, where child labor can be used for woodfuel collection and labor-intensive agricultural production (Dasgupta 1998). But large families become economically unsustainable in cities if family members cannot generate income in formal markets. While immigrants to urban areas initially retain rural traditions, including large families, the economic realities of urban life gradually reduce household sizes. Correspondingly, the demand for traditional fuels falls, both because of the rising time opportunity costs for urban residents working in formal markets, and because of the availability and economy of alternative energy sources (Chauvin 1981; Cline-Cole et al. 1990).

The data show that larger urban households tend to select traditional fuels to a great extent, whereas smaller households tend to choose more modern fuels (Table 3-1). Lack of time to start and maintain cooking fires is one reason that small families do not choose to cook much with traditional fuels. The choice is also correlated with income, inasmuch as large households in developing countries generally have relatively low per capita incomes and therefore low opportunity costs associated with the household labor time needed to utilize traditional fuels.

Although larger households generally choose traditional (and often less efficient) fuels in greater proportions, they generally consume less total energy per household member than do smaller households (see the rightmost column of Table 3-1). For most households in developing countries, cooking is the main end use for energy consumption, and the amount of energy required to cook for a large household is not proportionately greater than for a small household. Thus, all else remaining constant, the level of energy consumption per household member is smaller for larger families. This finding is consistent across the cities in the study, reflecting the importance of controlling for family size in any analysis of the urban energy transition.

Household Income

Urban income exerts a major influence on household energy choice. As incomes increase, there is a decline in the percentage of consumers who chose to consume traditional fuels (as well as the transitional fuel kerosene) and an increase in the percentage of consumers who chose to consume LPG and electricity (see Table 3-2). The trends for coal are unclear, except in the higher

TABLE 3-2. Relationship between Income and Energy Use in Urban Areas of 12 Developing Countries

Income class (per capita)	Monthly income US$/capita	Fuelwood	Charcoal	Coal	Kerosene	LPG	Electricity	Total
				Energy type				
				Fuel choice (% of households)				
Low	80.59	55.00	54.30	14.70	67.90	10.20	61.40	NA
Mid-low	15.51	38.70	44.00	17.00	62.80	22.40	70.10	NA
Middle	25.02	31.50	36.60	15.50	52.20	43.70	76.90	NA
Mid-high	41.94	26.10	37.30	4.70	40.70	59.80	79.30	NA
High	116.95	15.90	29.30	0.00	19.60	76.70	92.30	NA
				Fuel use (kgoe/capita/month)				
Low	8.59	3.63	3.28	2.38	1.33	0.15	0.60	11.59
Mid-low	15.51	2.57	2.66	3.21	1.73	0.42	0.82	11.59
Middle	25.02	2.10	2.20	2.83	1.50	1.25	1.15	11.15
Mid-high	41.94	2.62	2.54	0.67	1.14	2.09	1.77	10.82
High	116.95	1.66	1.79	0.00	0.60	3.70	4.15	11.62

Source: ESMAP Household Energy Surveys.

Energy used
(KgOE per person per month)

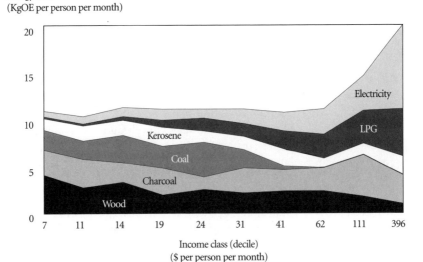

FIGURE 3-1. Income Class and Quantity of Fuels Consumed in 45 Cities

income class, reflecting the impact of the extensive distortion of that market in China (see Table 3-2).

Although income is strongly related to the energy type chosen, it is not as closely related to the total quantity of energy used, except in the higher income classes. Total consumption for the 80 percent of the population with low or moderate incomes is quite comparable (see Table 3-2 and Figure 3-1). The explanation lies in the fact that households shift from lower-efficiency traditional fuels to higher-energy-value modern fuels as they move up the income latter. People with higher incomes obtain more useful energy from their fuels. The relative consumption shares for wood, charcoal, and even kerosene also remain relatively stable with income for the 80 percent of the population with low to moderate incomes (see Figure 3-1).

Electricity and LPG consumption rise dramatically in the higher income groups (see Table 3-2 and Figure 3-1). Electricity demand in particular continues to increase with income due to continuing purchases of appliances. Indeed, the quantity of electricity used increases proportionally with income (Figure 3-2). In the highest-income households, the amounts of modern fuels purchased and consumed are substantially greater than those consumed in lower-income households.

The persistence of traditional fuel consumption beyond the lower income levels is surprisingly common for the countries in this study, encompassing households in African countries such as Burkina Faso, Mauritania, and Zambia, and high-income households in the cities of Port au Prince (Haiti),

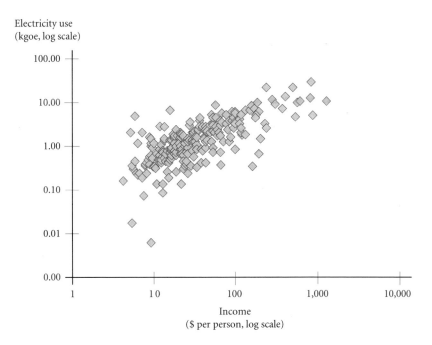

FIGURE 3-2. Electricity Use by Income in 45 Developing-Country Cities

Trinidad (Bolivia), Cagayan (Philippines), and Chiang Mai (Thailand). The explanation may lie in non-income-related and possibly location-specific factors such as access to wood supplies around cities, use of charcoal as a specialty fuel for grilling, and distorted market access to modern fuels.

It is also evident that poorer households use kerosene and, to a lesser extent, LPG. (Table 3-2 and Figure 3-1). As expected, LPG consumption rises with income, but the number of lower-income households that use LPG is still significant. Apparently, LPG is highly valued as a cooking fuel by those who can obtain it, regardless of income level.

Modern Fuel Access

While household variables strongly influence the urban household choices of fuel and consumption levels through the demand side, consumption behavior is also influenced by elements on the supply side. Access restrictions in the modern fuel sector are an important supply-side constraint that influences consumer behavior. The availability of modern fuels may depend on decisions by service companies to limit their distribution, government policies that affect fuel availability, and other factors.

Electricity

Service provision by electricity companies is an important constraint that influences a household's choice to use electricity. In our study, a qualitative index was constructed to measure access. The index reflects power company policies for extending service within urban areas, including the level of hookup fees, credit requirements, and service limitations (e.g., access limitations to a particular part of the city).

By this measure, urban access to electricity service is almost universal in some countries in the study, including Thailand and Bolivia.[6] In poor countries such as Mauritania and Burkina Faso, access to electricity in urban areas is significantly more limited (World Bank 1990c, 1991a). Access to electricity increases with the size of the urban area and with consumer income (see Table 3-3). Indeed, limited service access is the primary cause of the low electricity penetration in small towns in the poorer developing countries. When electricity is accessible, even low-income households in urban areas choose to use it. We thus observe a near-universal adoption of electricity in cities that do not do not have significant barriers for households to obtain a service connection.

Electricity access not only is associated with higher electricity use but also is inversely related to the use of traditional fuels (Figure 3-3). The availability of electricity in urban areas seems to act as a catalyst for people to switch from traditional to modern fuels. One possible explanation is that access serves as a proxy for market development. In that case, fewer barriers would constrain other modern fuels in cities where electricity is available. Another explanation

TABLE 3-3. Electricity Access and Fuel Use in 45 Cities

Electricity access	Income ($/capita/ month)	City size ('000)	Wood	Charcoal	Kerosene	LPG	Electricity
			Fuel choice (percentage)				
Very difficult	33	23	56.4	73.4	57.6	26.6	21.1
Difficult	67	174	72.3	33.5	65.2	21.8	42.8
Easy	62	514	24.1	62.7	50.4	21.6	47.7
Very easy	77	1,153	22.1	34.5	42.6	47.8	90.5
			Fuel use (kgoe/capita/month)				
Very difficult	33	23	1.31	10.09	0.35	1.49	0.24
Difficult	67	174	7.27	2.54	0.46	0.91	1.24
Easy	62	514	2.83	7.20	1.10	0.50	2.00
Very easy	77	1,153	1.71	1.75	1.75	2.00	2.79

Source: ESMAP Household Energy Surveys.

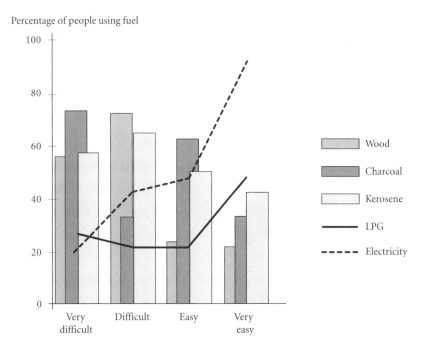

FIGURE 3-3. Electricity Access and Fuel Choice

would be that the availability of lighting and other appliances spurs people to a greater acceptance of modernity and modern fuels.

Zambia and Yemen illustrate the importance of access to electricity adoption (see Table 3-4). In Zambia, electricity charges are among the lowest in Africa, at about 2 cents per kilowatt-hour (kWh), and yet only 53 percent of urban households are connected to the electricity grid. The main reason for the low adoption rates in Zambia is restricted access. An example at the opposite extreme is the remarkably widespread use of electricity in the urban areas of Yemen, a country with relatively high electricity prices. The average price of electricity in Yemen is about 15 cents per kWh. Some 91 percent of urban households in Yemen use electricity. Price alone, therefore, is not a significant deterrent to the adoption of electricity.

Liquefied Petroleum Gas

LPG is an efficient, clean-burning, and highly valued cooking fuel. It can be ignited quickly (in contrast to woodfuel) and lacks the unpleasant smell associated with kerosene. The use of LPG can also reduce exposure to indoor air pollution. Women and children are the main beneficiaries of LPG adoption, because it saves labor time that might otherwise have been devoted to wood-

TABLE 3-4. Electricity Rates and Connection Policies in Main Urban Areas of 12 Countries

Country	Income $/capita/ month	City size ('000)	Electricity use and price % users	Kgoe/ capita	US¢/kWh	US$/kgoe	Kerosene price ratio
Bolivia	64	273	93	3.73	5.56	0.66	1.88
Burkina Faso	29	204	33	0.97	13.73	1.63	2.22
Cape Verde	54	54	66	1.08	20.30	2.41	4.26
China	18	59	100	0.62	4.30	0.51	1.52
Haiti	59	1,000	91	2.46	13.65	1.62	3.40
Indonesia	20	2,485	93	0.96	6.06	0.72	5.07
Mauritania	27	135	28	0.28	26.03	3.09	7.20
Philippines	32	2,067	93	1.53	5.98	0.71	2.82
Thailand	99	2,063	100	4.71	6.06	0.72	2.57
Yemen	70	272	91	2.28	14.99	1.78	3.37
Zambia	20	324	53	2.07	1.77	0.21	0.91
Zimbabwe	51	584	83	5.91	6.65	0.79	2.15

Source: ESMAP Household Energy Surveys.

fuel collection and/or the maintenance of cooking fires. Women and children also spend the most time around cook stoves and therefore receive the greatest share of the health benefits from cleaner cooking fuel.

Distribution restrictions in developing countries often constrain the use of LPG in urban areas. Government or private oil companies typically distribute LPG in 10 to 15 kg bottles. Consumers may have to purchase the bottles or make a sizable deposit. LPG is frequently distributed according to customer lists, and the lists are often restricted by ability to pay. Thus, households must reside in a relatively well-off income class to make the list and/or to have the resources to commit to the initial investment needed to adopt LPG. Chapter 6 offers a detailed case study of the impact of such access constraints on residential LPG utilization in Hyderabad India.

Effects of Pricing on Fuel Consumption

We now consider the effects of prices on energy utilization. In this assessment, the price of kerosene is used as the reference fuel to derive relative price ratios, that is, all fuel prices are standardized according to their ratios to the price of kerosene. The local price of kerosene is used as the reference fuel because of its status as a transition fuel and availability in all countries in the study. Using relative price ratios indicates the opportunity costs consumers face in their fuel choice and consumption decisions.

Price Effects on Woodfuel Consumption

High relative woodfuel prices induce fuel switching, usually towards charcoal. Figure 3-4 shows that when the absolute price of woodfuel is 18 percent of the price of kerosene, people will use woodfuel for more than 50 percent of their energy needs. But when the price of woodfuel is 60 percent higher than that of kerosene, woodfuel consumption drops to less than 5 percent of total consumption. Charcoal consumption expands to about 30 percent in this case, and the use of kerosene expands to just under 20 percent.

Thailand provides a historical example of the impact of relative woodfuel prices on energy use. In Bangkok, woodfuel was relatively inexpensive in the early 1980s, mainly due to logging activity in the country. However, the government banned logging in 1989 in response to high deforestation rates, and the prices of woodfuel and charcoal rose rapidly. In response, most people who cooked with woodfuel in Bangkok switched to modern fuels, although they continued to use charcoal for preparing specialty dishes. Woodfuel itself has all but disappeared as a cooking fuel in the major urban areas of Thailand.

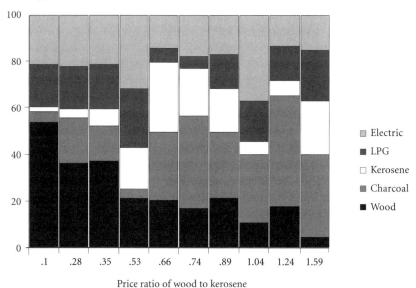

FIGURE 3-4. Use of Wood Declines and that of Charcoal Increases as Wood Prices Increase

Source: ESMAP Household Energy Surveys.

Price Effects on Transition Fuels: Charcoal and Kerosene

The relationship between charcoal prices and fuel consumption is more complicated than often assumed. The first complexity is that woodfuel and charcoal prices are interrelated. Both depend on local supplies of wood around urban areas, and these fuels are substitutable in consumption. In particular, charcoal often competes with woodfuel as a cooking fuel in urban areas where wood has become distant from urban centers, and woodfuel prices are relatively high. In that case, charcoal plays the role of a transition fuel. However, charcoal also competes with modern fuels in some end uses. Specifically, high-income households continue to use charcoal for specialty grilling after they have otherwise switched to modern fuels. In this case, charcoal plays the role of a de facto modern fuel.

Rising relative charcoal prices are associated with declining charcoal use. When the price of charcoal is about 75 percent of the price of kerosene, people use it for about 50 percent of their energy needs. However, when it is twice the level of kerosene, people use it for about only 10 to 15 percent of their energy needs and turn to woodfuel, kerosene, LPG, or electricity. The diversity of energy transitions discussed in the previous chapter at least partially reflects such relative price differences.

Haiti is one of the few countries for which time-series data are available for charcoal prices during a particular period. Forest cover in Haiti dramatically declined during the 1970s as a consequence of demand pressures associated with high population growth rates. By 1978, the total forest cover had shrunk to only 6.7 percent of land area (see Lewis and Coffey 1985; Stevenson 1989). During this period, the real price of charcoal in Port au Prince rose at an average annual compounded rate of just over 6 percent per year. In about 1988, the price of charcoal caught up with the backstop prices of LPG and kerosene. The prices of the three fuels were fairly competitive during the study period following (Figure 3-5). Although few poor people in Haiti used LPG or kerosene, mainly because of poor access to such fuels, the price of charcoal rose along with other commercial fuels. This rise was both a conse-

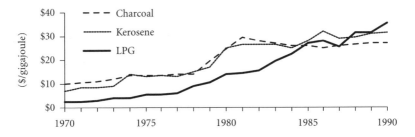

FIGURE 3-5. Energy Prices in Haiti, 1970–90
Source: World Bank (1991b).

Percentage share of energy use

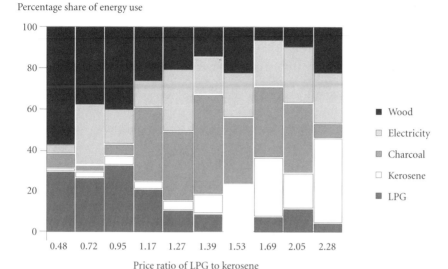

Price ratio of LPG to kerosene

FIGURE 3-6. Relationship of LPG Prices with Use of Five Fuels
Source: ESMAP Household Energy Surveys.

quence of the resource pressures noted, but also because taxes on kerosene and LPG displaced some demand to charcoal (World Bank 1991b).

Price Effects on LPG Consumption

The data indicate that LPG prices influence a household's consumption of LPG. Specifically, when the price ratio of LPG to kerosene is below one, households will use LPG for about 25 to 35 percent of their total energy consumption. As the price of LPG increases relative to kerosene, the use of LPG will decline to a lesser share in total energy use (Figure 3-6). The effects of relative prices on LPG consumption are evident in the experiences of Haiti and the Dominican Republic. Haiti is a country that historically taxed modern fuels, raising their prices. The retail price of LPG in Haiti averaged three times higher than in the Dominican Republic during the 1980s. The LPG market in the Dominican Republic was 12 times larger than in Haiti during this period (World Bank 1991b). The comparative price difference between the two countries for LPG fuel seems likely to explain at least part of this market asymmetry.

Price Effects on Electricity Consumption

The pattern of electricity consumption in developing countries is quite different than for LPG. Electric service is generally available in urban areas and used by most consumers, except for those living in poorer, more remote towns. The provision of electricity is typically viewed as a public service. In addition, urban consumers generally use electricity first for lighting, for which no other technology provides the same service quality.

The almost universal adoption of electricity in urban areas by residents who have access can be explained by the high value placed on electric lighting. The reason for this qualitative leap in energy services is the relative efficiency of electric lamps. An electric light bulb hangs from the ceiling and either fills the room with light or is focused downward for reading or close work. Indeed, a kerosene hurricane lantern gives off only about one-twentieth of the light from a 60-watt light bulb. Moreover, kerosene-generated light comes out of the side of the lantern, inevitably creating some glare that makes it less pleasant to use.

While the choice to use electricity is based on service access, income and relative prices determine consumption levels above a minimum threshold (again see Figure 3-2, and also Figure 3-7). Electricity consumption decreases with price to a greater extent in the highest income groups, while lower-income groups maintain the minimal service level. The poor can afford to use electricity for lighting, but they are slower to adopt other appliances, which (somewhat counterintuitively) makes their consumption behavior less price sensitive than higher-income customers (Westley 1992).

Consumption trends in the 1980s for Vientiane, Laos, provide an example of the way relative prices affect the level of electricity consumption. Households transitioned directly from biomass fuels to electricity in the period from 1983 to 1992 as a function of an expansion of hydroelectricity capacity, electricity availability, and low relative prices (World Bank 1993). The use of electricity soared as consumers not only began cooking with electricity but also began purchasing appliances such as small refrigerators and fans. People still use biomass for energy in Vientiane, but the growth of electricity use beyond the basic need for lighting has been dramatic.

Conclusions

The fuel choice and consumption decisions of urban energy consumers are affected by a combination of household attributes, incentives, and constraints. One of the strongest influences is household income, particularly with respect to the choice to use a fuel. The percentage of consumers who choose to consume modern fuels increases directly with income. The trend in

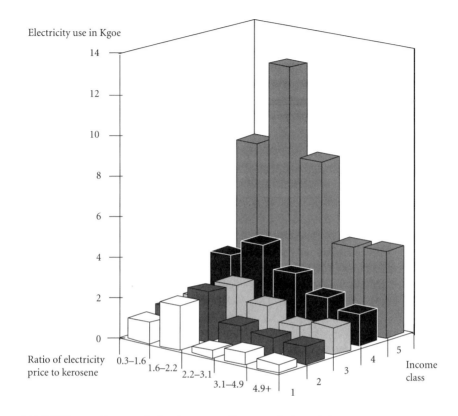

FIGURE 3-7. Energy Price, Income Class, and Electricity Use

Source: ESMAP Household Energy Surveys.

per capita consumption is not quite as direct. In this case, we see the lowest income classes consuming relatively large amounts of traditional fuels, while higher-income households consume a greater share of electricity and LPG. Between these extremes, fuel utilization patterns are influenced by family size, relative fuel access, and energy prices.

The results presented in this chapter provide an additional perspective as to why government policies appear to have played an influential role in urban energy markets. Incentive-based policies, such as taxes and subsidies, alter the energy prices urban consumers face. The evidence suggests that consumers are sensitive to relative energy prices. As a consequence, incentive-based policies are likely to shift the mix of fuels urban residents consume. The scope for government influence is likely to be particularly significant during the second stage of the urban energy transition, when consumers are in the process of changing their consumption behavior and when fuel alternatives are often available and price competitive.

The government has a particularly direct role in affecting the availability and price of modern fuels, both on the supply side, through infrastructure construction, and on the demand side, through access rationing, regulated pricing, and the use of taxes and subsidies. Such policies might be expected to affect the duration of individual stages and thus the pacing of the urban energy transition, as well as skew the consumption mix during stage two. For example, in Zambia, price controls on charcoal may have accelerated a consumption shift from woodfuel to charcoal. In Indonesia, kerosene subsidies seem likely to have accelerated the transition away from traditional fuels and to have acted as a disincentive for higher-middle-class families to switch to more modern fuels.

In summary, the evidence suggests that the fuel choice and consumption decisions of urban dwellers are sensitive to fuel access and energy prices, and that governments have used this fact in their policymaking. Perhaps it is encouraging that people respond to the incentives and choices they face. Fortunately or unfortunately, this means that policy intervention—whether of the "right" or the "wrong" kind—is likely to have consequences. It further suggests that policies need to be carefully crafted to achieve their intended effect and to avoid the unintended consequences.

4

Energy and Equity: The Social Impact of Energy Policies

Urban households in developing countries commonly face energy prob-lems that few city dwellers in industrial countries could imagine. Families may not be able to shift from dim kerosene lamps to high-quality electric lighting because the power utility does not serve their neighborhood. Households that would like to cook with clean and efficient LPG may face waiting lists of several years to obtain a service connection. Urban households that cook with fuelwood or charcoal may face seasonal price rises that stretch their budgets to the limit. Taxes may raise the price of biomass fuels to lower-income consumers.

This chapter focuses on the equity implications of the urban household energy transition and the impacts of sectoral policy on urban energy con-sumers. Household survey data are used to compare the quality of energy services, the level of energy consumption, and the fraction of energy expen-ditures in households of different income levels. The study also considers how energy policies affect poorer households in comparison with more well-to-do ones.

Several significant findings emerge from this assessment. Poorer urban consumers pay higher prices for useable energy due to the inefficiency of tra-ditional-fuel-using appliances and lighting fixtures. Poorer urban consumers also spend a larger share of their cash income on energy than do wealthier consumers. As the main users of traditional fuels, lower-income consumers also disproportionately bear the health and inconvenience costs associated with urban energy consumption.

Energy policy significantly affects the price of energy—hence, the percent-age of cash income that lower-income consumers dedicate to energy pur-chases. Energy policy options to improve the welfare of lower-income con-sumers include ending access inequity and instituting pricing reform in the

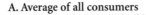

A. Average of all consumers

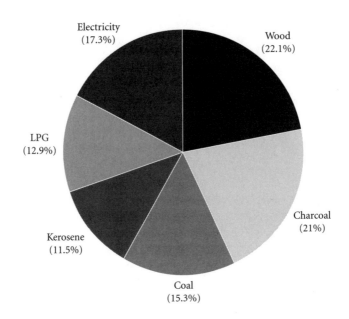

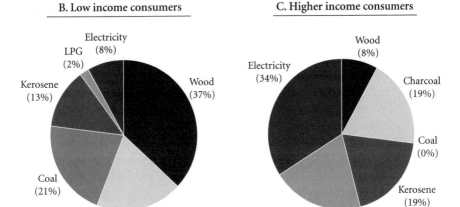

FIGURE 4-1. Household Energy Consumption in 45 Cities

Source: ESMAP Household Energy Surveys.

utility sector. Additionally, some policies, including taxes on transitional and modern fuels, should be avoided. The impact of structural economic reform policies that would affect urban energy markets need to be carefully evaluated, and they should be tailored to avoid adversely affecting the poor (see

Estache et al. 2001). The issues are not straightforward, however, and, as with other aspects of the problem of poverty in developing countries, simple solutions are seldom evident.

The chapter begins with a discussion of the relationship between energy consumption, expenditure, and income class. The next topic is the real price that lower-income classes pay for useable residential energy. We then examine the relationship between prices and fuel expenditures, before turning to the way income class affects the response of consumers to energy prices. The chapter concludes with a discussion of the policy implications.

Energy Consumption, Expenditures, and Income Class

Household consumption shares of the different fuel types, averaged across all income classes and countries, are shown in Figure 4-1 (panel A). Fuelwood and charcoal make up more than 40 percent of the energy consumed by urban households in the study. Kerosene accounts for only 10 percent of total residential energy consumption. Coal, which is specific to China as a residential fuel source, also accounts for about 10 percent of total consumption. Finally, the highest-value fuels—LPG and electricity—together account for about 33 percent of total consumption.

Differentiating consumption shares by income shows that the pattern of energy consumption is quite different for the poorest (lowest income decile in the sample; panel B) and wealthiest households (highest income decile; panel C). Woodfuel or charcoal accounts for more than one-half of the energy consumption of the poorest households in the population, while these fuels account for only one-quarter of the energy consumed by the wealthiest consumers.

The pattern of fuel expenditures differ from the pattern of energy consumption. Aggregating across income classes, the modern, high-quality fuels such as LPG and electricity dominate energy expenditures (see Figure 4-2). Expenditures on woodfuel and charcoal are about one-third of total money outlays on energy. The pattern reflects the fact that urban households in smaller and more remote cities may still be able to collect part of their woodfuel supply, thus lowering cash outlays, as well as the fact that customers are willing to pay relatively higher prices for the superior modern fuels.

Coal and kerosene do not account for a large percentage of consumer fuel expenditures in Figure 4-2 because consumer expenditures are reduced by substantial subsidies in the two countries in the study—China and Indonesia, respectively—where these fuels are consumed significantly. Together, electricity and biomass-based fuels comprise about three-quarters of all home energy expenditures.

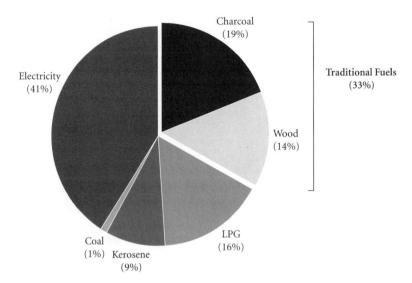

FIGURE 4-2. Average Household Expenditures on Six Fuels in 45 Cities
Source: ESMAP Household Energy Surveys.

It is sometimes thought that energy expenditures do not constitute a significant faction of poorer people's monetary income. However, enough of the
trade in traditional fuels is channeled through markets that energy prices do
affect consumer budgets. In fact, an analysis of expenditure data shows that
poorer people spend a significant share of their disposable cash income on
energy purchases. Families in the lowest income decile spend about $15 per
month on energy, while the higher-income families spend from $35 to $45
per month on energy (Figure 4-3). Translating these figures into relative
shares, the lowest income group spends more than 20 percent of its total
monthly cash income on fuel, the middle group spends about 12 percent, and
the highest income group spends just over 5 percent (Figure 4-4).

Even so, the 20 percent figure for the poorer households is likely to be
biased downward in terms of a complete economic accounting, because it
does not reflect the presence of barter trade, or the opportunity cost of labor
time dedicated to woodfuel collection and/or utilization, or the health costs
of indoor air pollution.

In summary, it is apparent that lower-income households are spending a
significant fraction of their cash income to obtain cooking energy, and they
are incurring significant nonmonetary costs as well. Consumption of these
traditional fuels does diminish in both relative and absolute importance with
rising incomes and increasing economic prosperity (see Chapter 3). But the
consumption of biomass-based fuels continue to dominate the consumption
mix and financial outlays for lower-income classes (see Peskin et al. 1992 for
the national accounting implications).

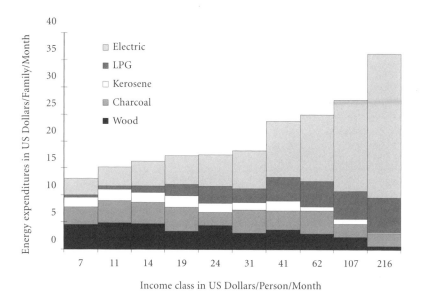

FIGURE 4-3. **Energy Expenditures by Income Decile in 45 Cities, 1988**

The Price of Poor People's Energy

It is often assumed that traditional fuels are relatively "inexpensive" vis-à-vis modern energy. As just discussed, it is clear that this assumption is not true if a full economic accounting is made of the cost of traditional fuels. Moreover, the market prices of residential fuels do not fully reflect consumer costs due to the fact that fuel-using appliances differ in their thermal efficiencies. This technical fact is not fully reflected in market prices in developing countries on account of incomplete information and other market imperfections in residential fuel markets (discussed below).

Beyond these accounting issues is the fact that the market prices of traditional fuels may themselves be relatively high compared with modern fuels. Supply shortages, sometimes seasonal in nature, frequently increase the market price of biomass-based fuels—particularly toward the later stages of the energy transition. As mentioned before, supply and distribution restrictions commonly limit access of poorer consumers to modern fuels, shifting excess demand to biomass-based fuels. Quantity rationing in modern fuel markets effectively dichotomizes the residential fuel market into modern and traditional segments. Under these conditions, modern and traditional fuel prices can be decoupled, and the market prices of traditional fuels can be higher than the market prices of modern fuels.

When all fuels are price- rather than quantity-rationed, market prices will be interlinked, although not identical, due to qualitative differences in fuel

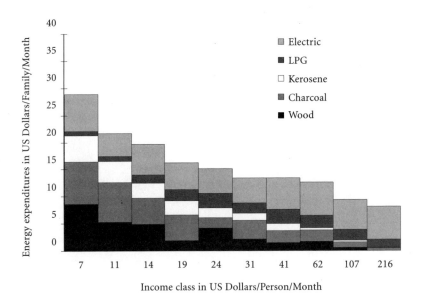

FIGURE 4-4. Percentages of Household Income Spent on Five Fuels, by Income Decile, 1988

properties. In this case, policies targeted at one fuel will influence the price of other fuels through substitution (see Chapter 3). This interlinkage has equity implications. Taxes on modern fuels will shift some demand to the traditional fuels that lower-income households disproportionately rely on, thus raising the price of energy used by these consumers and lengthening the time period that lower-income consumers will use traditional fuels. Conversely, subsidizing modern fuels will decrease the price of traditional fuels and accelerate the switch time from traditional to modern fuels. These trends will differentially affect poorer consumers who rely most heavily on traditional fuels.

The Price of Cooking Energy

The amount of heat that is burned for cooking is called the "input energy" and the amount that is actually absorbed by pots, pans, or other cooking vessels is called "useful" or "delivered" energy. The difference is waste heat that escapes around the sides of the pan. Lower-income households typically combust small quantities of fuels in thermally inefficient stoves or open fires with efficiencies (ratio of useful energy to input energy) ranging from 10 to 15 percent; charcoal stoves can reach efficiencies of up to about 25 percent. In contrast, LPG and electric stoves have thermal efficiencies varying between 55 and 70 percent (Fitzgerald et al. 1990). When adjustments are made for the thermal efficiencies of cooking appliances, it becomes apparent that the prices that lower-income households pay for cooking energy often are as high or higher than prices paid by more well-to-do households.[7]

To compute an income-class-specific price for cooking energy, we first adjusted market prices for the energy content of the fuels and the thermal efficiency of cooking stoves, yielding a price for delivered cooking energy for each of the different fuel types. Next, data on income-class-specific fuel shares were used to generate a weighted average price for each income class. From this information, we conclude that the poor pay more for delivered cooking energy than higher-income households in a number of countries in the study (see Table 4-1). These countries either have markets undistorted by policy interventions and the market price of traditional fuels is relatively high, or they have markets in which the price of traditional fuels has been raised directly or indirectly by taxation. The Philippines illustrates the former case. Poor consumers in the Philippines face relatively high prices for delivered cooking energy, especially in the smaller cities, where reliance on biomass energy is common. The difference between the delivered price of cooking energy for the income class in the top 20 percent and the lowest 20 percent in urban areas was close to $1 per kgoe at the time of the survey.

Haitian energy markets, distorted by taxes on petroleum products, further exacerbate barriers to access by the poor. The extremes in poverty and wealth in Haiti mean that costly petroleum products are used by only a few of the wealthiest households, displacing much of the demand for the rest of the income spectrum to charcoal, itself relatively expensive and inefficiently combusted.

TABLE 4-1. Countries in Which the Poor Pay More for Energy than Do the Wealthy

		$/kgoe of energy weighted by fuel use and efficiency					
		Income class					
		One (lowest income)	Two	Three	Four	Five (highest income)	Average of income classes
Bolivia	Undistorted Market	...	0.66	0.65	0.59	0.55	0.60
Cape Verde	Undistorted Market	1.55	1.67	1.62	1.58	1.49	1.58
Haiti	Tax	1.60	1.45	1.42	1.37	1.26	1.42
Philippines	Undistorted Market	1.79	1.48	1.24	1.01	0.66	1.36
Thailand	Undistorted Market	...	...	1.03	0.92	0.86	0.90
Yemen	Undistorted Market	1.79	1.89	1.37	1.45	1.22	1.45

Note: Ellipses indicate that the lowest income quartile(s) of these cities fall above the lowest quartile(s) for all cities in the study.

Bolivia follows a relatively undistorted market pricing regime and, as an oil producing country, the price of modern fuels is close to the world market price. Both the rich and the poor consume petroleum products. Relative to the other countries in which the poor pay more for energy, the differences between rich and poor households are noticeable but relatively modest. The difference between the price of delivered energy for the highest and lowest 20 percent of the urban population is only 9 cents per kgoe, reflecting the fact that petroleum-based fuels are cost competitive with fuelwood in this particular country.

In some countries, lower-income households pay about the same price for delivered cooking energy as do higher-income classes. This relative equality occurs in countries that either subsidize modern fuels in price-rationed markets or tax modern fuels in dichotomized quantity-rationed markets. Energy policy in China illustrates the first case. A coal subsidy in urban areas induces downward price pressure on all fuels, which are substitutable and generally accessible to all income classes. Except in one city in the study, lower-income groups pay about the same price as higher-income groups, and almost everyone uses coal for both heating and cooking. The exceptional case is one poor county where coal availability is limited. This county is located between two mountain ranges, and it is difficult to truck coal to the county seat. As a consequence, the energy market is dichotomized, and poor families with below-average incomes rely on wood, which is not subsidized. They pay more for delivered cooking energy than people with higher incomes.

In Indonesia, subsidized kerosene is widely available to all income classes in just about all markets (World Bank 1990b). Although the delivered price of wood for cooking is higher than that for kerosene, there is little difference between the income-class-weighted prices of cooking fuels in Indonesia due to the relatively low share of biomass-based fuels consumed by any income class in the country. Policies followed in Burkina Faso and Mauritania reflect the taxation case. Modern fuel markets are relatively undeveloped, and modern fuels are generally not available to the poor in these countries. However, the policy of taxing LPG and keeping electricity rates high effectively raises the delivered price of fuel for more well-to-do households to the level of the prices of charcoal and fuelwood used by lower-income consumers (Table 4-2). In these very poor countries, both higher- and lower-income households are paying high prices for useable cooking energy.

The Relationship Between the Delivered Price of Cooking Energy and Energy Expenditures

This section assesses the relationship between the delivered price of cooking energy and money outlays on cooking energy consumption. Figure 4-5 shows that energy prices and expenditures are strongly, though not perfectly, corre-

TABLE 4-2. Countries in Which the Poor Pay the Same Prices as Do the Wealthy

		$/kgoe of energy weighted by fuel use and efficiency					
		Income class					
Country	*Policy*	*One* *(lowest income)*	*Two*	*Three*	*Four*	*Five* *(highest income)*	*Average of income classes*
Burkina Faso	Tax	1.93	1.86	2.00	1.78	2.21	1.92
China	Subsidy	0.23	0.17	0.17	0.12	...	0.19
Indonesia	Subsidy	0.49	0.47	0.47	0.47	...	0.48
Mauritania	Tax	2.22	2.04	1.81	1.76	1.95	1.97
Zambia	Subsidy	0.64	0.64	0.60	0.55	...	0.62

Note: Ellipses indicate that the highest income quartile(s) of these cities fall below the highest quartile(s) for all cities in the study.

lated in the countries in the study. Countries on the left side of Figure 4-5 face relatively low prices for delivered cooking energy, while the countries on the right face relatively high prices.

Delivered cooking energy is cheapest for consumers in China and Indonesia, where consumers extensively rely on subsidized transition fuels (coal and kerosene, respectively). The percentage of income spent by both rich and poor urban consumers is relatively low in both countries. At the opposite end of the spectrum, the delivered price of cooking fuel and energy expenditures are relatively high in Haiti, Cape Verde, Burkina Faso, Yemen, and Mauritania. These countries have few indigenous energy resources (or, in the case of Haiti, have over-exploited the indigenous resources), and/or they tax modern fuels such as LPG. Combined with the shortages of indigenous biomass fuels, taxes make delivered cooking energy very expensive across all income classes, and energy expenditures are relatively high. The prices of woodfuel- or charcoal-generated energy in these countries are sometimes higher than modern fuel alternatives—the result of inefficient fuel markets.

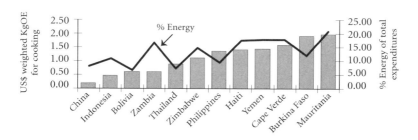

FIGURE 4-5. Weighted Energy Price and Percentage of Expenditures on Energy for Urban Areas in 12 Countries

Countries with available biomass resources and/or undistorted energy markets fall between these extremes. Bolivia has the lowest prices for residential energy, mainly because the country produces LPG and makes it widely available. The Philippines and Thailand pursue similar market-oriented strategies, and people in these countries consequently also enjoy low energy prices—though not as low as in the countries that subsidize fuels. Thus, as expected, the policy to tax or subsidize a fuel used for cooking does have an effect on the percentage of expenditures that households utilize to pay for their energy bills.

The Price of Lighting

Lighting is the first use of electricity in urban households. Kerosene is used either when electricity is not available or during power outages. The efficiency of electrical lighting is much higher than that of kerosene lamps or candles. In fact, it is estimated that one 60-watt light bulb has an equivalent amount of light output as 18 kerosene wick lamps (Nieuwenhout et al. 1998; van der Plas and de Graaff 1988) and uses much less energy to produce the light. Florescent lights are even more efficient, reaching efficiency levels more than 200 times greater than kerosene lamps (Nieuwenhout et al. 1998).

Electrification affects not only people's total energy use but also lifestyle and quality of life. Here a distinction needs to be made between cooking and lighting. People tend to use lighting more extensively once they acquire an electrical hookup, whereas they generally do not change the amount of food they cook. The superiority of electric lighting often induces households to expand and diversify their activities when they acquire electricity, using lighting for reading, sewing, and recreation.

Our analysis of electricity prices is based on the concept of lighting service. Due to the superiority of electricity, we assumed that households with access to electricity use it as the primary source of household lighting, while those without electric hookups use kerosene. A weighted average price for lighting services was then computed for each income class, based on the percentage of households in each class using electricity and kerosene. This assessment shows that countries in the study with the most inequitable lighting costs are Haiti, Cape Verde, Mauritania, and Burkina Faso, mainly because the lower-income classes in these countries do not have access to electricity, and because kerosene prices are high (Table 4-3). Poorer residents, who must use kerosene, pay higher prices for lighting services than do higher-income customers. In Indonesia and the Philippines, lower-income households also pay more for lighting services than do higher-income households, but the price difference between the two is only a matter of 3 to 8 cents per kilolumen hour (a kilolumen hour is the quantitative measure of lighting service). In addition, the poorer households pay about one-fifth less for lighting services in

TABLE 4-3. Price of Lighting by Energy Type and Income Class in 45 Urban Areas

	Kerosene and electricity lighting prices (US$/kilolumen hour)							
	Prices by type		*Combined kerosene and electricity prices by income class*					
	Electricity	*Kerosene*	*One*	*Two*	*Three*	*Four*	*Five*	*Average*
Countries in which the poor pay more than do the wealthy								
Burkina Faso	0.52	2.45	2.07	1.98	1.85	1.44	1.17	1.82
Cape Verde	0.77	1.90	1.40	1.30	1.32	1.13	0.85	1.17
Haiti	0.52	1.65	0.71	0.60	0.58	0.57	0.56	0.60
Indonesia	0.23	0.47	0.26	0.25	0.23	0.23	...	0.25
Mauritania	0.99	1.49	1.57	1.50	1.31	1.25	1.12	1.41
Philippines	0.23	0.87	0.34	0.26	0.24	0.23	0.26	0.27
Countries in which the poor pay about the same as do the wealthy								
Bolivia	0.21	1.15	...	0.34	0.34	0.26	0.24	0.28
China	0.16	1.17	0.19	0.17	0.15	0.17	...	0.17
Thailand	0.23	0.94	...	...	0.21	0.23	0.24	0.23
Yemen	0.57	1.83	0.74	0.67	0.66	0.64	0.64	0.65
Zambia	0.07	0.77	0.40	0.40	0.38	0.42	...	0.40
Zimbabwe	0.25	1.27	0.47	0.42	0.41	0.52	0.39	0.44
Average	0.37	1.23	0.68	0.64	0.65	0.57	0.43	0.61

Note: Ellipses indicate that the income quartiles of these cities was not available for the study.
Source: ESMAP Household Energy Surveys.

both Indonesia and the Philippines, compared with the first group of countries.

In other countries, electricity is universally available, and lighting is widely available at the market price for all income classes. The urban poor pay about the same as other urban residents for lighting in China, Bolivia, and Yemen, because almost everyone in the urban areas of these countries has access to electricity service. The exceptions to this pattern are in Zambia and Zimbabwe, countries in which urban electrification plans are based on a regional coverage approach. In these countries, the regions that gain electricity contain both lower income and more wealthy households, and most people in the regions adopt electricity regardless of whether they are rich or poor. The regions without electricity also contain people with many different income levels. The regional coverage approach is more equitable than providing service to only those that can afford higher levels of electricity service, but it does deny access to consumers in noncovered regions.

Another exception to the trend is in Thailand, where the poor actually pay less for lighting. The reason is that almost everyone in major urban areas uses electricity, and the government has a policy to provide "lifeline" rates to households that use very little electricity. These lifeline rates reduce the price that the poor pay for electricity service to less than the price paid by more wealthy households. Thus, for all countries in the study, the main factor affecting the different prices that poorer or wealthier households pay for lighting is access to electricity.

Income Class and the Response to Energy Prices

The poor and the middle classes are responsive to price signals for traditional fuels, as distinct from more wealthy households that typically do not significantly consume traditional fuels. The fact that approximately two-thirds of urban populations are responsive to traditional fuel prices means that higher prices of traditional fuels will lead to fuel switching and/or energy conservation. Lower-income households will be affected by changing traditional fuel prices, whether these changes are induced by the government—through subsidizing or taxing transition fuels—or by market conditions.

The demand for traditional energy and kerosene by lower-income households should be quite price elastic when all these fuels are available, due to relatively close substitutability among them and to the large impact of energy prices on the budgets of low-income households. When fuelwood, charcoal, and kerosene are not all available, consumers will be less sensitive to price, given that cooking energy demand is difficult to economize. For the same reason, LPG consumption is likely to be relatively price insensitive for the high-income consumers who use it as a cooking fuel. The demand for electricity is not very price elastic for low-income classes, who use electricity mostly for lighting, and it is more elastic for upper-income classes who have the flexibility to economize on the use of luxury appliances such as air conditioners (see Figure 3-7).

Because lower-income households do not use large quantities of modern fuels, one might conclude that changes in the price of kerosene, LPG, or electricity will not much affect them, and that governments are therefore relatively free to institute policies in the modern fuel sector with little consideration of the distributional side effects. This conclusion, however, would be erroneous on two counts. First, poor people in many countries do use electricity for basic lighting services and are thereby directly affected by price changes for electricity. Secondly, price changes in modern fuel markets affect the poor through demand displacement and substitution to traditional fuels by the higher income classes, if modern and traditional fuel markets are not decoupled by quantity rationing.

Conclusions and Policy Implications

The evidence suggests that low-income urban households rely on traditional fuels to a greater extent than do higher-income households, and that lower-income households are paying higher prices for useable energy due to the inefficiency of traditional-fuel-using cooking stoves and kerosene lamps. Lower-income consumers also spend a larger share of their cash income on energy than do wealthier urban consumers. As the principal consumers of traditional fuels, lower-income consumers also bear a disproportionate share of the health and inconvenience costs associated with residential energy consumption. Taken together, these findings suggest that the poor are relatively burdened by the pattern of residential energy utilization in developing countries.

One important policy variable is fuel access. Expanding fuel access will increase the flexibility to substitute one fuel for another, and it will increase the responsiveness of lower-income consumers to the price of different fuels. This flexibility could reduce inconvenience and monetary costs, allowing lower-income households to lessen the burden of energy taxes on traditional or transitional fuels and to face lower monopoly markups in monopolistically competitive markets. The flexibility to fuel switch also allows consumers to minimize the burden of price increases caused by local shortages. The prices of woodfuel and charcoal in developing countries often rise during the rainy season. If the poor lack access to fuel alternatives—through limited supplies, economic barriers to fuel access, or institutional restraints—they have no choice but to pay relatively high prices in a restricted traditional fuel market. Policies to promote market access include investments both in modern fuel infrastructure and distribution, and in reducing the financial barriers to entry for the poor. The barriers can be reduced for the poor by providing loans for stoves or through rolling up-front costs into the price of energy. For instance, the price of LPG or electricity could be raised very slightly to cover the cost of initiating new service.

The price of lighting is generally the consequence of electrification policy. Both the poor and the rich are able to take advantage of electrical lighting in countries that promote universal service. Where barriers to electricity adoption exist, the poor are usually differentially restricted and pay more for lighting services. One potentially sensible energy assistance program for the poor that is carried out by many developing-country power companies is the practice of using increasing block-rate tariff structures along with connection charges that are rolled into the overall price that the public pays for electricity. These practices reduce the barrier to entry and allow the poor to adopt electricity for the low levels of electricity they use—about 50 kilowatt hours per month. In countries with such policies, most of the poor take advantage of electricity service for lighting. The level of service should be geared to this use level, and it should be accompanied by wide access to connections, which will forestall the common practice among the poor of buying electricity from a neighbor, often at higher prices than they would be charged by the power distribution companies. This practice is more prevalent in urban areas of developing countries than many power distribution companies realize. Thus, the adoption of the block-rate tariffs potentially could benefit poor customers as well as the cash flow of utility companies.

Broadly applied subsidies on transitional or modern fuels would also help lower-income households. The effects of such subsidies are both direct and indirect. The direct effect is that many poor people use the transition fuel, thereby reducing their energy expenditures. The indirect effect is that biomass fuels must compete with transition fuels that are subsidized and widely available to the poor. Therefore, subsidizing substitute fuels provides a price cap on traditional fuels. Many countries with high biomass fuel prices have large markups between roadside supply of the fuel and final market price in urban areas. This margin often contracts with changing supply costs and greater competition from fuel alternatives.

Broad-based subsidies on modern fuels benefit all income classes. In fact, the better-off classes will garner the greatest share of modern fuel subsidies because they consume more modern energy than do lower-income households. Subsidies also impose budgetary and economic costs and foreign exchange costs if subsidies encourage additional fuel importation. Despite these negatives, it should be reemphasized that fuel subsidies do help the poor. In addition, foreign exchange costs would not be notable for the countries in this study because the main subsidies are for indigenously produced fuels—coal in China and kerosene in Indonesia. Indeed, subsidizing import-competing domestic fuel sources could actually reduce foreign exchange costs, thereby providing an additional benefit.

To overcome the drawbacks of general subsidies, some countries have tried limiting fuel subsidies to rationed quantities. This policy avoids some of the problems of more broadly applied subsidies, but it is difficult to apply and administer effectively (see Chapter 6). Moreover, the price differential between subsidized and unsubsidized fuel provides a motive for those who control fuel distribution to divert shipments onto the open market. The market bifurcation between subsidized and unsubsidized fuel also has the drawback of not capping the price of traditional fuels at the lower subsidized level. In short, the direct benefit of this kind of subsidy policy can be attenuated, and the indirect benefit of the general subsidy—lowering the backstop price of traditional fuels—is absent.

Either subsidizing modern fuels or taxing traditional fuels could be used as a policy instrument to accelerate the energy transition for environmental, health, or other reasons. However, subsidizing modern fuels would have a positive effect on the poor, while taxing traditional fuels would have a negative effect in a market where poor consumers are already burdened. For this reason, using policy to accelerate or subsidize the modern fuel sector would seem the better of the two options. Again, however, the practical problems of subsidies, as well as their budgetary impact and net economic cost, would have to be weighed in the decisionmaking.

Sometimes kerosene, LPG, and electricity are taxed in developing countries. The intention is often laudable—taxing "rich people's" fuel is a progressive way to generate revenue and encourage energy conservation. In the end, however, such taxation hurts the poor by raising the effective cap that modern fuel prices put on biomass fuels. Not only does this policy hurt the poor in the short term, it delays the energy transition in the long term by raising the backstop price of traditional fuels.

Another point is that structural economic reform proposals that would remove subsidies in the energy sector or deregulate prices (e.g., for electricity) must be assessed very carefully (Estache et al. 2001). If subsidies are removed without a compensatory program, lower-income households will be hurt the most. Unfortunately, many countries cannot continue to provide subsidies for financial reasons, because residential energy demands are growing more rapidly than energy demands in other sectors, making household subsidies an increasingly untenable financial burden.

In the past three chapters, we have been focusing on energy evolution and utilization patterns in urban energy markets, consumer attributes affecting consumption demand and the responses of consumers to incentives and constraints, and the particular issues that confront low-income consumers in urban energy markets. Throughout these chapters, we have considered the role of policy. In the next chapter, we shift the focus to consider the environmental and health implications of urbanization and traditional fuel consumption patterns. We specifically examine the relationships between urbanization, residential energy consumption, and periurban deforestation, and the implications of biomass consumption trends on the exposure risk consumers face to the indoor pollutants generated from using biomass fuels for cooking.

5

The Urban Energy Transition and the Environment

More than 25 years ago, Eric Eckholm (1975) focused the attention of the international development community on the "other energy crisis." At the time, the world was facing oil shortages, sharply rising petroleum prices, and long queues at gasoline stations. Forecasters were predicting that fuel prices would continue to rise, and the United States began filling the strategic petroleum reserve. But Eckholm contended that policymakers were ignoring an equally serious problem: looming shortfalls of biomass energy in the developing world. His projection was based on Club of Rome–type extrapolations in which population growth would proportionately increase the consumption of woodfuels, and biomass stocks would decline in proportion to consumption (Agarwal 1986; Eckholm 1975). This market dynamic would result in unsustainable tree harvesting, a "woodfuel gap" between excess demand and supply, and significant deforestation.

A number of researchers in the energy field criticized the "woodfuels gap" hypothesis for its absence of economic content, that is, for failing to allow for the possibility of endogenous supply augmentation, household energy conservation, or interfuel substitution (Cline-Coal et al. 1990; Dewees 1989; Foley 1987; Mercer and Soussan 1990). A more recent body of literature has emerged on the so-called "environmental Kuznets" curve, which is conceptually consistent with the critique of the woodfuels gap hypothesis (see Dasgupta et al. 2002; Panayotou 2003; Stern and Commons 2001). In the environmental Kuznets curve scenario, resource utilization increases during an initial period of economic development, and associated health and environmental effects worsen. This stage is consistent with a Club of Rome/Eckholm-type forecast—health and environmental problems increase roughly in proportion to population growth during the early economic growth period. But unlike Eckholm's predictions, the environmental Kuznets curve scenario asserts that this "scale effect" gradually is mitigated over time through a number of possible adjustment responses: consumer demand shifts towards more environmentally benign commodities, as a consequence of changing preferences or rising incomes; fuel switching occurs in response to rising prices; technical efficiency experiences gains in resource utilization; and more stringent environmental policy is adopted. Beyond a certain income threshold, the sum of the mitigating effects begins to dominate the scale effect of economic expansion, and the environmental impacts of economic growth begin to diminish.

The conceptual framework of the Kuznets curve literature can readily be extended to the concept of the urban energy transition. An environmental Kuznets curve in this context would imply enough substitution to modern fuels after a certain urban growth threshold to begin to reduce the adverse effects of biomass fuel consumption on human health and the environment. In fact, extending the logic of the Kuznets curve concept to the natural limit, the question could be raised: what is the urban growth threshold beyond which cooking energy consumption becomes entirely disassociated from periurban deforestation and the health effects of biomass-based fuel burning—the state reached in advanced economies—due to comprehensive fuel switching?

The woodfuels gap scenario and the environmental Kuznets curve extension bracket a range of possible outcomes in terms of the projected resource and environmental futures associated with urban energy transitions. And as a first approximation, it would appear that the actual reality since Eckholm's original warning falls somewhere in the middle of the range. Just as the world has recovered from the energy crisis (at least temporarily), predictions that the developing world would run out of biomass energy appear to have been overstated. At least the time frame of the projections has not fully materialized. But as pointed out in Chapter 4, the price of traditional fuelwood in urban markets is sometimes very high—higher than that of access-restricted modern fuels, in fact—reflecting relative resource scarcity. Deforestation also continues to be regarded as a significant problem locally around many cities. But perhaps most significantly, a rapidly growing body of literature on indoor air pollution provides evidence that the health effects from biomass-based cooking emissions is one of the most significant health issues in the developing world (see Smith 2003; Smith and Mehta 2003). While a significant part of the problem lies in rural areas, studies show that the urban poor, who continue to rely on biomass-based cooking fuels, are also at risk.

This chapter examines the relationship between urban growth and the environment to assess the resource, environmental, and health effects of the urban energy transition. We first investigate the relationship between urban income growth and local deforestation around cities. We then examine how demand pressures on periurban resources through the biomass consumption channel evolve as cities grow and energy markets develop. The subject next turns to the implications of fuel consumption trends on indoor air pollution exposure risks. Lastly, the chapter concludes with a summary of the results and a discussion of policy implications.

Periurban Forestation Patterns

What is the relationship between urban income growth and forestation patterns around cities? To address this question, we assess the degree to which standing biomass in the vicinity of cities corresponds to a number of supply and demand variables. The study is based on a subsample of 34 cities (in the countries marked "a" and "c" in Table 1-1).[8] Data limitations ultimately narrowed the focus of the analysis to the following variables: distance from urban areas, urban income, periurban population density, transportation development, and two natural environmental factors (topography and precipitation). The averages of these variables, aggregated at the country level, are provided in Table 5-1. Annex 2 describes the methodology and sources for the derivation of the variables in Table 5-1.

TABLE 5-1. Characteristics of the Sample Countries and Regions

Country/ region	Natural biomass range (m³/ha)	Mean periurban biomass (m³/ha)	Mean city income (10⁶US)	Mean periurban roads (km/10⁶ha)	Mean periurban topography[a]	Mean periurban rainfall (mm/yr)
Indonesia	88–252	29	1,279	753	0.96	1,689
Philippines	8–345	40	1,114	664	1	1,976
Thailand	78–120	30	2,331	473	0.34	1,320
Asia average	NA	33	1,575	630	0.77	1,661
Botswana	0–71	30	62	275	0	466
Mauritania	0–32	2.6	32	228	0	227
Zimbabwe	16–103	48	186	431	0.67	769
Africa average	NA	20.2	77	317	.17	454
Bolivia	15,242	76	344	373	1.48	545
Haiti	80,300	11	489	571	2	1,500

Note:

[a] Topography figures are based on dummy variables that assumed the following values:

0 = Flat plains with relatively small topographic change; < 500 m elevation change over a 25 km horizontal distance (2 percent slope).

1 = Moderately sloped areas with gently rolling hills; 500 to 2,000 m elevation change over 25 km horizontal distance (2 to 8 percent slope).

2 = Elevation changes greater than 2,000 m over 25 km horizontal distance (> 8 percent slope).

NA = not applicable

The measure of standing biomass used in the study is average biomass density (m³/ha). The first step for determining its value was to obtain vegetation maps displaying the spatial distribution of different vegetation classes around each of the cities. The distribution outlines of different vegetative classifications were digitized and plotted using an ATLAS*DRAW/ATLAS* Graphics software package—a quasi-geographic information system (GIS) package—producing detailed computerized maps of the vegetation patterns in concentric zones around each city. Figure 5-1 shows the results for Surabaya, Indonesia, as an example. The software was then used to compute the areal extent of each vegetation class, and vegetation-class-specific "biomass conversion factors" were applied to convert the areal calculations into standing biomass estimates associated with each vegetation class. The next step was to aggregate biomass estimates for each vegetation class to yield the total standing biomass in each zone. This figure was then divided by the land area in each zone to yield zonal biomass density values. This process allowed us to generate detailed mapping of standing biomass densities in concentric zones around each city we studied.

Biomass Distribution around Cities. We now turn to patterns in the data with respect to biomass densities in periurban regions. As expected, biomass densities around cities in the study tend to increase with distance from urban areas. Averaging across cities within each country, the average standing biomass density in zone 2 (26–50 km from city centers) is greater than that in zone 1 (0–25 km from city centers) in six out of eight countries in the study (see Table 5-2). There are generalizable differences among countries in this periurban forestation pattern. The difference in standing biomass stock between the proximate (0–25 km) and outlaying zones (26–100 km) is greatest for Haiti and the Asian countries—Indonesia, the Philippines, and Thailand. Average standing biomass for Haiti in zone 2 is 3.2 times higher than in zone 1, and 2.1, 3.8, 4.2 times higher respectively for Indonesia, the Philippines, and Thailand. The discrepancy in city-averaged biomass densities between the inner and outer periurban zones is not as pronounced in the African countries and Bolivia. For example, the second zones in Mauritania and Zimbabwe have biomass densities averaging 1.8 and 1.3 times higher, respectively, than the inner-most urban zones. City-averaged biomass density in zone 2 in Botswana is actually 0.97 of the biomass in the innermost zone; in Bolivia, the figure is 0.87.

These differences can be explained by the interplay of human factors that induce deforestation and natural factors that influence the level of standing biomass. We now turn to an assessment of both groups of factors.

City Income and Periurban Population Density. City income and periurban population density are two important human factors that induce defor-

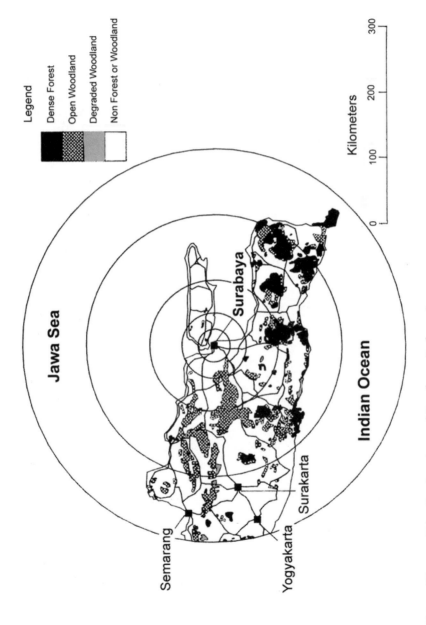

FIGURE 5-1. Biomass Distribution Around Surabaya, Indonesia

TABLE 5-2. Values for Biomass, Population, and Other Variables in Periurban Zones

	Zone	Distance (km)	Haiti	Indonesia	Philippines	Thailand
Biomass density	Z1	0–25	4.96	13.30	15.05	6.20
(m³/ha)	Z2	26–50	15.96	28.14	57.23	25.98
	Z3	51–100	12.97	31.78	56.71	29.61
		Ratio Z2/Z1	3.22	2.12	3.80	4.19
Population density	Z1	0–25	3.57	15.35	22.30	7.92
(persons/ha)	Z2	26–50	0.03	0.64	0.35	0.10
	Z3	51–100	0.02	1.16	0.20	0.86
		Ratio Z2/Z1	0.01	0.04	0.02	0.01
Road density	Z1	0–25	720.00	715.50	772.60	682.33
(km/10⁶ ha)	Z2	26–50	536.00	361.67	629.60	356.33
	Z3	51–100	457.00	414.17	589.40	321.33
		Ratio Z2/Z1	0.74	0.51	0.81	0.52
Topographic relief	Z1	0–25	2.00	0.33	0.20	0.00
(topography index)	Z2	26–50	2.00	0.83	1.40	0.33
	Z3	51–100	2.00	1.67	1.40	0.33
		Ratio Z2/Z1	1.00	2.50	7.00	—
Precipitation	Z1	0–25	1533.00	1243.17	—	1369.33
(mm/year)	Z2	26–50	1500.00	2093.67	—	1394.33
	Z3	51–100	1467.00	1840.50	—	1360.33
		Ratio Z2/Z1	0.98	1.68	—	1.02

	Zone	Distance (km)	Mauritania	Botswana	Zimbabwe	Bolivia
Biomass density	Z1	0–25	1.88	31.26	41.10	78.60
(m³/ha)	Z2	26–50	3.44	30.32	52.73	68.58
	Z3	51–100	2.82	29.76	54.38	78.20
		Ratio Z2/Z1	1.83	0.97	1.28	0.87
Population density	Z1	0–25	0.87	0.37	1.96	1.09
(persons/ha)	Z2	26–50	0.01	0.03	0.04	0.02
	Z3	51–100	0.00	0.03	0.02	0.02
		Ratio Z2/Z1	0.01	0.08	0.02	0.02
Road density	Z1	0–25	402.20	447.67	670.75	459.60
(km/10^6 ha)	Z2	26–50	195.00	207.67	330.67	262.40
	Z3	51–100	87.40	191.67	305.67	155.20
		Ratio Z2/Z1	0.48	0.46	0.49	0.57
Topographic relief	Z1	0–25	0.00	0.00	0.50	1.00
(topography index)	Z2	26–50	0.00	0.00	1.00	1.40
	Z3	51–100	0.00	0.33	0.67	1.40
		Ratio Z2/Z1	—	—	2.00	1.40
Precipitation	Z1	0–25	237.00	466.00	797.00	761.40
(mm/year)	Z2	26–50	227.20	463.00	812.67	806.00
	Z3	51–100	243.50	476.33	833.33	894.00
		Ratio Z2/Z1	0.96	0.99	1.02	1.06

estation around cities. The size and income of cities, and population densities around them, will stimulate land conversion and biomass energy demand in the periurban region.

City income and population densities differ widely among the cities in the study, and they are likely to be an important determinant of the periurban forestation patterns observed. Average city incomes are significantly higher for the Asian cities than for Haiti, Bolivia, and especially the African cities (see Table 5-1). Averaged population densities in the 0–25 km zones follow the same pattern. For example, average population densities in zone 1 (measured in persons per hectare) range between 7.92 in Thailand to 22.30 in the Philippines; the figure is 3.57 for Haiti (see Table 5-2). In contrast, average population densities in zone 1 around Mauritania and Botswana are only 0.87 and 0.37 respectively, with the values for Zimbabwe and Bolivia at 1.96 and 1.09 respectively.

These patterns, and their inverse relationship to periurban biomass distribution (see Table 5-2) suggest that urban income and population density could play a role as causative deforestation agents. Supporting this conjecture is the fact that average standing biomass density within 100 km of the Asian cities in the sample is about 33 m³/ha, about 50 percent higher than the African average of 20 m³/ha (see Table 5-1), while natural standing biomass densities in Asia are often 10 to 30 times higher than in African regions in the sample. These patterns suggest that some combination of urban income and population growth has passed a threshold in Haiti and the Asian countries to be able to significantly affect forest resources in the proximate regions to city centers.

The effect of urban growth, periruban population density, and distance on periruban deforestation is also suggested within Asia by comparing the standing biomass around cities of different sizes. Comparing the two Philippine cities of Manila and the much smaller Cagayan de Oro, for example, shows that biomass densities successively decline around both cities moving outward from zones 1 to 3. However, biomass density in every zone around Manila is less than that of Cagayan de Oro (Figure 5-2).

Transportation Development

Transportation development is also correlated with urban income growth and population density. The expansion of transportation infrastructure increases the accessibility of biomass stocks to urban expropriators, and it acts as an inducement for settlements, which are likely to intensify along transportation corridors. These populations may also place pressure on forest stocks.

The impact of transportation development is likely to be another reason that land around Port au Prince, Haiti, and the Asian cities in the study has

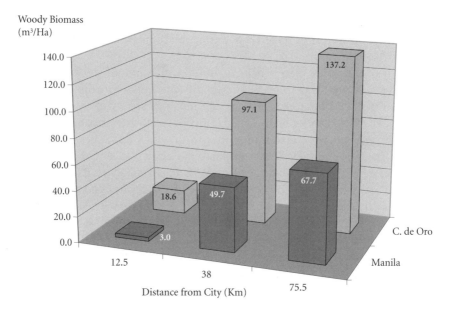

FIGURE 5-2. Standing Biomass around Cagayan de Oro and Manila, Philippines

been heavily deforested. The average road density within 100 km of Asian cities in the sample is 630 km/10^6 ha (see Table 5-1). The figure for Port au Prince is also very high, about 571 km/10^6 ha. The African average is only 317 km/10^6 ha; for Bolivia, the figure is about 373 km/10^6 ha. In all countries in the study, transportation networks are more developed in the most proximate zone around cities (see Table 5-2). Because transportation development and population density are both inversely correlated with distance from cities, it is not necessarily distance as such—a proxy perhaps, for "pure transportation cost"—that reduces deforestation pressures in the hinterlands. More sparsely developed transportation networks and lower population densities may also lower resource demands in more distant regions away from urban centers.

Natural Variables

Natural variables can also be expected to interact with the human demands on periruban resources to influence the spatial distribution of forest resources around cities. The topography around periurban areas and local precipitation patterns would seem likely to be two particularly important factors influencing forestation patterns.

Topography. Topographical variation might be postulated to have two countervailing effects on local forest resources. First, the remoteness and inaccessibility of periurban regions in more mountainous terrain should increase harvesting costs, when all else is constant, thus reducing deforestation pressures. On the other hand, more mountainous terrain would tend to concentrate periurban populations in more accessible low-land areas, increasing local demand pressures there. These potentially countervailing effects are hard to disentangle.

Topography appears to have some influence on the forestation patterns around cities in the study in Asia, and possibly Haiti and Bolivia. These counties exhibit a greater degree of topographic relief than do the African countries in the sample, with the exception of Zimbabwe (see Table 5-2). Haiti provides the interesting case of exhibiting high topographic relief across all periurban zones. By concentrating population pressure in lowland areas, topography in Haiti might account for some of the contrast in average standing biomass with other countries in the study, such as the countries in Africa, where topography does not appear to significantly affect forestation patterns. On the other hand, because topography is relatively constant across zones in Haiti, it cannot not explain the diminished biomass densities between the innermost and outer zones. In this case, human factors such as population density and road density are likely to be responsible. The Haitian context differs from Asia, where topographic relief averaged across cities increases at greater distances from cities (again see Table 5-2). In contrast to Haiti, topography in Asia seems likely to interact with human factors to determine the level of standing biomass at different distances from urban centers.

A comparison of the Indonesian cities of Bandung and Surabaya illustrates the possible influence of topography on forestation patterns in Asia (Figure 5-3). These cities have nearly the same populations and incomes, yet biomass density within the first two zones around Bandung is more than 75 m³/ha, whereas biomass density for Surabaya is only 6 m³/ha within 25 km of the city and about 18 m³/ha in the 26–50 km zone. Topography is likely the principal explanation. Bandung is situated in a mountainous region, whereas the first two zones around Surabaya are in regions without much topographic relief. The comparative remoteness of forest resources around Bandung in this case likely protects forest resources around this city from human resource demands.

With the exception of Bandung in Indonesia and Cagayan de Oro in the Philippines, topography in all Asian cities of the study is relatively flat within 25 km of cities. However, five of the eight Asian sample cities have hills and varied terrain in the 26–50 km zone. In the outermost zone, only Bangkok, located in a river delta, still is surrounded by relatively flat land. This finding suggests that topographic patterns likely interact with human factors to explain the pattern of standing forest around Asian cities. The relatively flat

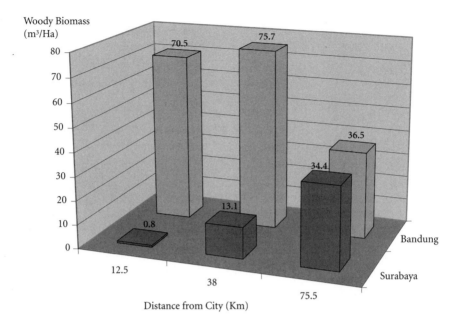

FIGURE 5-3. Biomass Supply around Bandung and Surabaya, Indonesia

terrain around Bangkok is likely to partially explain why this city has the least standing biomass in the outlying zones of any of the Asian cities in the sample.

Precipitation and Other Variables. Precipitation and other natural variables, such as soil conditions, temperature, and geology influence the biomass distribution around cities. At the level of country averages, precipitation is not a confounding variable with periruban distance in our study, because precipitation is relatively constant across zones around cities (see Table 5-2). On the other hand, the differences across countries in the level of precipitation (Table 5-1) may partially explain between-country differences in natural biomass densities. Natural biomass densities are positively correlated with the level of precipitation (see Table 5-1).

Piercing beneath the veil of the country averages, we can see the influence of precipitation on biomass densities in particular cities. In Mutare, Zimbabwe, standing biomass drops from about 62 m³/ha to 54 m³/ha moving from zone 1 to zone 2, about a 14 percent decline (see Figure 5-4). Precipitation declines by 28 percent between these zones, and vegetation shifts from open montane forest (62 m³/ha) in zone 1 to a mix in zone 2 that includes dry bushy savanna (33 m³/ha) and wood grassland (16 m³/ha). The

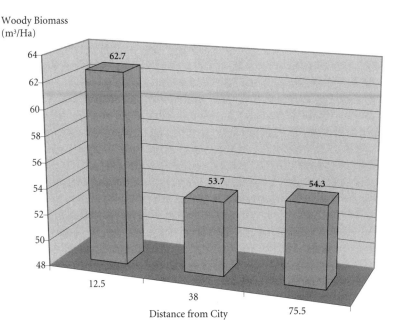

FIGURE 5-4. Standing Biomass around Mutare, Zimbabwe

changes in the natural variables in this case are sufficient to obscure the impact of any human-induced deforestation activities in the vicinity.

Natural factors are also important determinants of the biomass pattern observed around cities in Bolivia. For example, virtually no standing biomass can be found around Oruro and La Paz within 25 km of the cities (Figure 5-5). The biomass level for La Paz increases in the 26–50 km zone to about 24 m³/ha, whereas for Oruro it remains close to the zero level. This result might seem counterintuitive, because the population of La Paz is five times greater than that of Oruro. The pattern is likely to be explained by the fact that while both cities are located at high elevation above timber line, zone 2 for La Paz is partially within a lower elevation region that contains humid montane forests having biomass densities of 126 m³/ha. Zone 2 for Oruro, on the other hand, remains largely above timberline.

The Potential Role of Urban Energy Consumption

The previous section showed how a combination of human and natural factors interact to influence the spatial distribution of forests around cities. We now focus more specifically on the potential role urban energy consumption itself could play in placing demand pressures on periruban forest stocks.

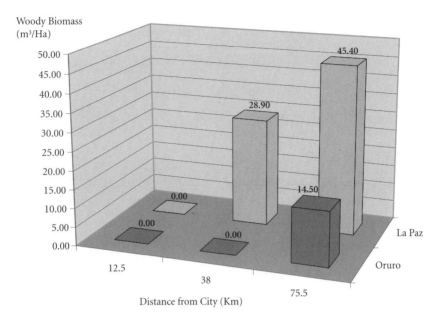

FIGURE 5-5. BIOMASS AROUND LA PAZ AND ORURO, BOLIVIA

Once urban residents move out of extreme poverty, the per capita con-sumption of traditional fuels (woodfuel plus charcoal) persists at a relatively stable level until the higher income levels (see Table 3-2 and Figure 3-1). Fuel switching to modern fuels does not significantly occur until cities grow in size beyond the one million population level and a significant upper middle class develops (Table 2-1). It is only for the wealthier and larger cities in our sam-ple that the per capita consumption of traditional fuels is declining.

However, even when traditional fuel consumption declines on a per capita basis, the aggregate consumption level might continue to increase with grow-ing urban populations due to the scale effect of economic expansion. Aggregate consumption is the better measure of the impact of urban energy utilization on biomass stocks around cities, because it is directly related to aggregate resource demand. Table 5-3 ranks cities by aggregate biomass con-sumption, from highest to lowest (column 1). The sample is subdivided approximately into thirds; the subtotals and averages refer to these groupings.

A number of interesting points arise. First, each grouping has cities in all transitional stages of the energy transition. In fact, the top four cities in terms of aggregate biomass consumption include one city in stage 3 (Bangkok) and three cities in stage 2 (Lusaka, Port Au Prince, and Nouackchott). It is not until the fifth-ranked city is reached—Ouagadougou—that stage 1 is encoun-tered. Thus, it is certainly not the case that cities that pass out of stage 1—the stage in which biomass consumption dominates—necessarily evince a decline in aggregate biomass consumption.

However, there is some relationship between energy transition stage and aggregate biomass consumption. The average transition rank of the top third of the cities is 1.8, the middle third is 2.1, and the bottom third is 2.3 (last column of Table 5-3). These figures indicate an inverse relationship between energy transition stage and aggregate biomass consumption. However, it is not completely clear how strongly this correlation would hold if city size were held constant. A direct relationship does appear to exist between the size of cities and their aggregate biomass consumption. With the exception of La Paz and Oruro, the bottom cohort tends to be the smallest group of cities in the study—as well as the cities relatively more advanced (as a cohort average) along the energy transition. And the top group of cities tends to be the larger cities, which are also the cities less further along the energy transition. The corresponding city size rank (second to last column in Table 5-2) averages 10.5, 14.5, and 26.8 (out of 34) for the top, middle, and bottom group of cities respectively.

These trends show that the per capita consumption of biomass fuels can be relatively low, and yet aggregate consumption can be relatively high. In such cities as Bangkok and Manila, populations are large enough to generate substantial aggregate resource demand, even with relatively low per capita biomass consumption. In this case, declines in aggregate biomass energy consumption can lag behind the per capita trends, and the demand pressures on surrounding forested land can continue even after cities have reached the later stages of the modern fuel transition.

Another striking fact is the increase in the number of cities in stage 2 moving from the bottom to the top cohort; in fact, the majority of cities in the top two cohorts are in Stage 2. The following stylized scenario seems consistent with this and the previous observations. In the "general case," cities start out small, and biomass is the dominant fuel. As they grow in size and economically develop, consumers begin to fuel switch to the transitional fuels, and the per capita consumption of biomass fuels ultimately declines. But the aggregate level of biomass consumption continues to rise for a significant part of stage 2 because the correlated city size expansion initially dominates any decline in per capita biomass fuels consumption. This statement would hold less strongly for less rapidly growing cities and more strongly for more rapidly growing cities. In the latter case, it would be possible to generate substantial aggregate demand for biomass energy even in stage 3, if the scale effect of the population increase continues to dominate the per capita decline in biomass consumption. Bangkok is an illustrative case in the present study.

We further note that a number of researchers have documented a relationship between poverty and population growth (e.g., Dasgupta 1999), and that Chapter 4 documents a relationship between poverty and the propensity for consumers to consume biomass-based fuels. Putting these factors together suggests that aggregate biomass consumption is likely to remain higher for a longer period of time in poorer, rapidly growing cities than in wealthier, slower growing cities. A city such as Ouagadougou would illustrate the first case (rank number 5 for aggregate biomass consumption); a city such as Quillacollo (rank number 32) would illustrate the second.

Another important point to note is the relative increase in the share of charcoal in aggregate biomass consumption as one moves from the bottom to the top group of cities. Charcoal accounts for 66 percent of aggregate biomass consumption in the top third of the cities; it accounts for 41 percent in the middle third, and 38 percent in the bottom third (see "percent biomass" column in Table 5-3). The data reflect the importance of charcoal as a stage 2 transition fuel and the fact that the cities in stage 2A of the urban energy transition—the stage marked by a switch from woodfuel to charcoal—dominate in the high biomass consumption group. It is clear that cities that follow the charcoal-dominated transition path to modern energy usage will impose relatively large demands on the resource base during the transition period.

In this regard, it is instructive to compare Jakarta, the largest city in the study, and Manila, the second largest. Although Jakarta's population was 2 percent larger than Manila's at the time of the energy surveys, the aggregate biomass consumption for Manila (ranked 6th) was 2.8 times higher than for Jakarta (ranked 13th), and the aggregate charcoal consumption for Manila was 4.1 times higher. Note that Manila is in transitional stage 2C—the stage in which undistorted market policies allow a relatively diversified fuel consumption mix—while Jakarta is in transitional stage 2B, a stage in which kerosene is a dominate transition fuel. Although a number of factors could partially explain this difference, it seems likely that the kerosene subsidy in Indonesia has had the effect of both reducing the total aggregate demand for biomass-based fuels and reducing particularly the demand for charcoal as a transitional fuel. Presumably, the difference in the consumption profiles between Jakarta and Manila would have been even more extreme had Manila followed the stage 2A transition path, that is, provided incentives for charcoal consumption. In any event, the actual difference in biomass consumption associated with the stage 2B and 2C transition paths actually followed by Jakarta and Manila suggests that the impact on periurban resources from cooking energy during the stage 2 transition has been substantially more significant in Manila than in Jakarta.

Implications for Indoor Air Pollution Exposure Risk

Like the "energy crisis," which was first conceived of as a problem for developed economies, concerns about indoor air pollution first gained attention in

TABLE 5-3. Rank Order of Cities by Aggregate Biomass Consumption

Rank	Total biomass (KGOE/Mo)	Wood	% Biomass	Charcoal	% Biomass	City size rank*	Energy stage[a]
1 **Lusaka**	8626	1061.4	0.12	7564.6	0.88	10	2A
2 **Port Au Prince**	8227	0.0	0.00	8227.0	1.00	8	2A
3 Bangkok	5548	193.7	0.03	5354.3	0.97	3	3
4 Nouakchott	4866	127.1	0.03	4738.9	0.97	13	2A
5 Ouagadougou	4538	4520.8	1.00	17.2	0.00	14	1
6 Manila	4513	1925.8	0.43	2587.2	0.57	1	2C
7 Kitwe	4355	1504.7	0.35	2850.3	0.65	16	2A
8 Davao	2713	2346.2	0.86	366.8	0.14	9	1
9 Surakarta	2381	1891.6	0.79	489.4	0.21	11	2B
10 Bobo Dioulasso	2364	2197.1	0.93	166.9	0.07	18	1
11 Yogyakarta	2205	1529.5	0.69	675.5	0.31	12	2B
Subtotals	**50336**	**17298.0**	*	**33038.0**	*	*	*
Averages	**4576**	**1573.0**	**0.34**	**3003.0**	**0.66**	**10.5**	**1.8**
12 Luanshya	1862	496.9	0.27	1365.1	0.73	23	2A
13 Jakarta	1625	1000.8	0.62	624.2	0.38	2	2B
14 Sanaa	1485	1205.9	0.81	279.1	0.19	15	3
15 Livingstone	1389	596.5	0.43	792.5	0.57	24	1
16 Semarang	1154	989.6	0.86	164.4	0.14	6	2B
17 Bandung	1004	608.4	0.61	395.6	0.39	4	2B
18 Cagayan de Oro	837	802.3	0.96	34.7	0.04	17	1

	Zone	Distance (km)	Mauritania	Botswana	Zimbabwe	Bolivia
Biomass density (m³/ha)	Z1	0–25	1.88	31.26	41.10	78.60
	Z2	26–50	3.44	30.32	52.73	68.58
	Z3	51–100	2.82	29.76	54.38	78.20
		Ratio Z2/Z1	1.83	0.97	1.28	0.87
Population density (persons/ha)	Z1	0–25	0.87	0.37	1.96	1.09
	Z2	26–50	0.01	0.03	0.04	0.02
	Z3	51–100	0.00	0.03	0.02	0.02
		Ratio Z2/Z1	0.01	0.08	0.02	0.02
Road density (km/10^6 ha)	Z1	0–25	402.20	447.67	670.75	459.60
	Z2	26–50	195.00	207.67	330.67	262.40
	Z3	51–100	87.40	191.67	305.67	155.20
		Ratio Z2/Z1	0.48	0.46	0.49	0.57
Topographic relief (topography index)	Z1	0–25	0.00	0.00	0.50	1.00
	Z2	26–50	0.00	0.00	1.00	1.40
	Z3	51–100	0.00	0.33	0.67	1.40
		Ratio Z2/Z1	—	—	2.00	1.40
Precipitation (mm/year)	Z1	0–25	237.00	466.00	797.00	761.40
	Z2	26–50	227.20	463.00	812.67	806.00
	Z3	51–100	243.50	476.33	833.33	894.00
		Ratio Z2/Z1	0.96	0.99	1.02	1.06

the developed world. Indeed, indoor air pollution concerns were at first linked to the developed world's energy supply problems. In the post-oil-embargo period of the late 1970s, the Carter administration launched a set of policy proposals designed to conserve energy in residences and commercial buildings. "Tightening up the building shell" was among the conservation options available for reducing building energy consumption, but it soon became evident that reducing the air exchange in buildings not only conserved energy but also had the potential to increase the concentration of indoor air pollutants. Because the majority of people spend most of their time indoors, the quality of indoor air they breathe has obvious health implications. Moreover, the quality of indoor air had not been controlled for in previous studies assessing the health impacts of outdoor air, raising questions about the accuracy of the risk assessments in those studies.

While indoor air quality is still a pertinent concern today in developed countries, the attention of researchers has shifted also to the possible health effects of indoor air pollution in developing countries. In fact, indoor air pollution is now seen as a top health priority of international development agencies such as the World Health Organization (WHO) of the United Nations. Specifically, biomass-based cooking fires produce high emissions rates of carbon monoxide (CO), total suspended particulates (TSP), nitrogen dioxide (NO_2), and polycyclic aromatic hydrocarbons (PAH), due to poorly controlled combustion—a common condition in developing countries. Inadequate venting to the outside and poor circulation translate emissions rates into high indoor air concentrations. The concentrations of combustion-generated pollutants in the indoor air of households in developing countries are sometimes much higher than WHO standards and national guidelines (Smith 2002). And the proximity of cooking fires to women and children increases their exposure risks by orders of magnitudes, compared with pollution generated from outside sources (Smith 2003; Smith and Mehta 2003; Smith et al. 2000).

Of particular concern in the literature is the contribution of indoor air pollution to the incidence of acute respiratory infections (ARI). Young children living in homes in which biomass fuels are used for cooking are two to three times more likely to contract ARI than unexposed children, when other factors are controlled for (Smith 2002). Other respiratory ailments, such as chronic bronchitis, are also linked to indoor air pollution from cooking. Women cooking around biomass fires for a number of years are two to four times more at risk from chronic obstructive pulmonary disease (COPD) than unexposed women (Smith 2002). Indoor air pollution poses a number of other possible health effects, such as asthma. Health effects research is now very active in the indoor air pollution field.

What are the implications of our data with respect to indoor air pollution exposure risks? There would appear to be several. First, moving out of poverty—increasing income beyond the very lowest income class—would appear to reduce indoor air pollution exposure risk, all else constant, because a decline in the per capita consumption of both woodfuel and charcoal is observed when households move above the lowest income point (see Table 3-2 and Figure 3-1). Beyond that point, however, the per capita consumption of biomass-based fuels (woodfuel plus charcoal) is relatively stable until the highest income levels (again see Table 3-2 and Figure 3-1), suggesting that there is a relatively wide range of incomes for which some groups of urban residents will experience indoor health pollution health effects for a relatively long period of time. Public policies and resource conditions are likely to cause some variation in exposure risk within this middle income range, however. Middle-income consumers in cities with substantial surrounding biomass resources or government policies that impede the development of the modern fuel sector—or discourage cleaner-burning transitional fuels, like kerosene—are likely to have the highest exposure risk from indoor air pollution. Contrariwise, residents of cities with limited biomass resources, or those subject to policies that encourage a more rapid transition to clean transitional fuels or modern fuels, will face lower exposure risks to indoor air pollution.

The possibility of relatively high aggregate biomass consumption particularly in stage 2 cities (see Table 5-3), in conjunction with relative constancy in per capita consumption trends with city size and income until the highest levels (see Tables 2-1 and 3-2) suggests that the total population exposed to indoor health pollution risk does not decline significantly until the very end of the energy transition. However, this aggregate risk level may be mitigated by the use of charcoal, which often poses problems for regional deforestation but burns cleaner than does woodfuel. Again, variations from this generalization likely reflect periurban resource conditions, public policy, and income distribution. Cities with larger low-income populations, more abundant periurban biomass resources, and government policies that discourage modern fuels use are most likely to have sizeable populations that experience indoor air pollution risks for a long period of time.

Finally, it is worth noting the special case of China, a country in which the lower and middle income classes generally use coal as cooking fuel. Indoor coal combustion poses health risks to occupants that will persist unless ventilation systems are satisfactory or until a complete transition to modern fuels is made (Lan et al. 2002).

Conclusions and Policy Implications

As a general pattern, biomass stocks around urban areas are inversely correlated with population densities and transportation development in and around cities, and they are directly correlated with distance from the urban center—up to around 50 km. However, location-specific factors are likely to cause a unique expression of the deforestation pattern around each urban area. Natural environmental variables such as topographic variation and precipitation patterns are the factors most likely to inject irregularities into what would otherwise be a smooth expansion of the deforestation perimeter as cities expand and economically develop.

The data suggest that deforestation has likely been more pronounced around cities in Asia and Haiti than in Africa and Bolivia, in the sense that variation in standing biomass attributable to human action stands out more clearly above the background variation caused by natural factors in these regions. Even in Asia, though, the interaction of urban expansion and natural variables seems evident. A striking example is a comparison of Bandung and Surabaya, two Indonesian cities of comparable size and economic developmental status, but with greatly different levels of periurban standing biomass. The relatively abundant resources around Bandung are attributable to mountainous terrain, which reduces resource accessibility.

Urban energy-driven pressures on the forest base also can continue as cities pass through the urban energy transition. We surmise that aggregate biomass consumption is likely to be particularly high from the end of stage 1 through the end of stage 2, especially for cities that follow the charcoal-dominated stage 2A transition route. Given the number of larger cities in stage 2 that account for the most aggregate biomass consumption, it is evident that per capita declines in biomass-based fuel consumption are not sufficient to compensate for the scale effect of expanding urban populations. The scale effect may dominate even after cities have passed through the energy transition to modern fuels. Bangkok is a city in stage 3 of the energy transition, yet it has the third largest aggregate consumption of biomass fuels of the 34 cities in the study.

These facts have serious implications for indoor air pollution exposure risks. Indeed, if anything, the continuing risk of indoor air pollution exposure through stage 2 and into stage 3 is greater than the risk that biomass fuel consumption will continue to impose pressure on periurban forested land. The reason is that there are many pressures on forested land besides the demand for biomass fuel—it is therefore not presumptively obvious what the total impact of biomass fuel consumption is relative to the sum total of all deforestation forces. Moreover, biomass fuel demand can be met by supply expansion. In contrast, the link between indoor air pollution exposure risk and biomass-based cooking is direct. Unless residents change habits (for example, keep children away from cooking fires), use more efficient stoves, or improve ventilation systems, continuing use of biomass fuels will lead to continuing exposure risks.

Although a longitudinal study would be needed to reach a definitive conclusion, the evidence in our study does not suggest the existence of an environmental Kuznets curve in which aggregate biomass consumption declines after a particular urban growth threshold. At the very least, no city in the study has reached the stage of cities in advanced economies, where the use of biomass cooking energy is negligible and thus where cooking energy utilization itself imposes almost no resource demand on periurban forests nor creates any biomass combustion-related exposure risk in indoor air.

Because urban growth itself does not seem likely to reduce exposure risks of indoor air pollution in the developing world at any time in the near future (nor does it diminish deforestation pressures), policy intervention has an important role to play. Policies to accelerate the urban energy transition would reduce cumulative indoor air pollution exposure risks that urban dwellers sustain over their lifetime, and they would reduce resource demands on the urban perimeter during urbanization. Even in relatively developed larger cities such as Bangkok and Manila, encouraging interfuel substitution (away from charcoal use in particular) would reduce indoor air pollution exposure risks and deforestation pressures. Crafting effective and low-cost policies to change the source of cooking energy for lower-income urban dwellers is a particularly crucial challenge policymakers confront, given the relatively high consumption of biomass-based cooking fuel by this class of consumers.

6

The Energy Transition in Hyderabad, India: A Case Study

Past chapters have exploited the information contained in comparative analyses of urban energy consumption and periurban resource impacts in a large sample of cities in Africa, Asia, Latin America and the Caribbean, and the Middle East. This chapter departs from the cross-sectional approach to consider a specific household energy survey conducted in 1994 in Hyderabad, India.[9] Trends in urban energy consumption and biomass supplies are documented and assessed in this survey. The Hyderabad case manifests many of the patterns and results previously discussed; for example, the role of government policies, resource constraints, and income disparities in influencing fuel choice and consumption levels, and the effects of evolving urban energy patterns on biomass stocks (Alam et al. 1998). The 1994 Hyderabad results are also compared with those obtained in a study conducted for Hyderabad City during the period of 1981–1982 (Alam et al. 1985b; Bowonder 1987a; Bowonder 1988) and with other urban studies in India as well (Alam et al. 1985a; Nair and Krishnaya 1985; Reddy and Reddy 1983, 1984). This longitudinal comparison provides additional insight into the way in which the factors previously analyzed influence the urban energy transition.

Interfuel substitution for urban households is taking place rapidly in Hyderabad. Much of the growth in the consumption of modern fuels has come from the liberalization of energy markets and the increased availability of such fuels as LPG, along with the increasing purchasing power of middle- and higher-income groups. It is likely that this trend will continue and even accelerate during the coming decades.

The structure and composition of domestic energy consumption in metropolitan Hyderabad is assessed in this chapter, with particular emphasis on the extent of the transition from traditional to modern fuels over the past decade. The impact of this transition on fuelwood demand and its effects on surrounding rural areas is also evaluated.

Energy Policies and Programs Affecting Urban Hyderabad

Because the markets for energy in urban Hyderabad have been strongly influenced by policies of the national government, we begin this chapter with an overview of the main policies influencing the urban energy sector. These policies include household fuel subsidy programs for petroleum products, limits on the import of petroleum products, and electricity distribution and pricing policies.

Kerosene Subsidies through Ration Card Program

Kerosene has long been viewed as a poor person's fuel in urban India, and the government has utilized a public distribution system for promoting the sale of kerosene products. Consumers are issued ration cards that cover a variety of products, including basic foodstuffs such as rice, as well as kerosene. A ration card permits the holder to purchase the covered products from designated retailers called ration shops. Kerosene is available at the ration shops in limited quantities at subsidized prices. In theory, kerosene in India is sold only through ration shops; other kerosene sales are illegal. Although the ration card system is helping poor people afford kerosene, it is not well targeted—it also benefits more middle-class and wealthy households, both of which have access to ration cards. In fact, the poorest households without addresses cannot obtain ration cards and are paying world market prices for kerosene, or they are using very expensive fuelwood. Complicating matters further, the ration shops often receive kerosene only periodically, so the availability of the fuel is sporadic.

LPG Distribution through Government-Affiliated Retailers

LPG has traditionally been distributed through retail dealers associated with the national petroleum companies. Until the 1990s, these distribution companies had exclusive rights to sell LPG, but they were required to follow pricing policies set by the government. The pricing policies were based on the principle that LPG should be sold at a price that reflected internal production costs in India, regardless of the world market price. However, there was not enough LPG produced within India to satisfy fuel demand. As a consequence, a system was developed to limit the number of families that could purchase LPG from the distribution companies. The LPG retailers developed a customer list, limited the number of LPG bottles customers could have, and serviced only customers on that list. Retailers have tended to concentrate on established—and typically well-off—customers because their supplies are insufficient to meet the total demand. Other customers are put on a waiting list. Because LPG is a highly desirable cooking fuel and fairly inexpensive compared to wood and other fuels, tens of thousands of people had put their names on the waiting list to obtain LPG. More recently, with the liberalization of LPG markets and the importation of LPG by government petroleum companies, the waiting list for public LPG has become much smaller.

Recent policies have opened up the LPG market to private retailers, but they are permitted to sell only imported LPG. The price of LPG obtained from these retailers is higher than the price charged by the government-affiliated retailers. The combination of an increased number of private retailers and an expansion of LPG supply to government-affiliated retailers has resulted in a tremendous growth in LPG use in Hyderabad. The increasing LPG supplies have meant that the market for this product has expanded among middle- and high-income consumers, and even some poor consumers. It also has meant that the annual subsidy going to LPG users in Hyderabad at the time of the 1994 survey was approximately US$10 million in Hyderabad alone, and about US$500 million for India as a whole. This figure has risen to well over US$1 billion today. The Hyderabad study finds that the middle class and the wealthy appreciate the convenience of LPG and will continue to use it even with higher prices. Also, the switch to LPG has had the benefit of freeing up kerosene for the urban poor.

Electricity and the State Electricity Boards

In most of India, State Electricity Boards supply the electricity to urban households. The State Electricity Board is a vertically integrated government monopoly that controls production, transmission, and distribution of electricity to urban households. Although urban households receive some subsidies as a result of electricity pricing policies, the main subsidies in the electricity sector go to rural agricultural consumers. The State Electricity Boards are suffering financial strain because of these agricultural subsidies, and at the same time they are under great pressure to provide better service to urban consumers. By trying to maintain near-universal service levels with limited financial resources, the quality of service has declined significantly, including many brownouts, blackouts, and voltage drops. The public opinion in Hyderabad toward the State Electricity Board is quite negative. Focus group interviews of middle-class and higher-income households reveal support for reforming and liberalizing the distribution of electricity, even if it means higher prices. There is not a similar level of support from poor people, which indicates that special consideration should be given to addressing their problems in the event of market reforms.

Fuelwood as Sole Market-Based Household Fuel

Whereas modern fuels are heavily regulated, the fuelwood trade in Hyderabad is based on market principles. The main trend for fuelwood has been a shift from serving household customers to serving commercial ones. In the space of nearly 15 years, the number of people using fuelwood as a cooking fuel has declined significantly, so the market has contracted and reoriented itself toward other customers. Many commercial customers are using wood to replace or to enhance the poor-quality coal previously used in their enterprises. Despite continued use by the commercial sector, deforestation in the Hyderabad region has slowed considerably as a result of the consumption shift away from household use. Existing policies should be strengthened to maintain this important trend.

To summarize, government policies have been extremely important in shaping household energy demand in Hyderabad. As will be discussed later, the energy consumption subsidies in urban Hyderabad are quite large (more than 80 million rupees [Rs] per month) and are likely to continue to grow as more people switch to modern fuels. Unfortunately, these subsidies, which are meant to help the poor, are not very well directed. In fact, the majority of the subsidies end up in middle- and upper-class households that can afford to pay market rates for fuels. Moreover, policies to limit imports of fuels create periodic local scarcities. In such scarcity situations, it is the poor who are most disadvantaged, as evidenced by their lack of access to kerosene until LPG became more widely available to middle-class households.

The Transition to Modern Fuels in Hyderabad

In this section, we examine the overall patterns of energy use in Hyderabad, along with the changes in energy use in urban households. We evaluate the structure and composition of energy consumption, with particular emphasis on the transition from traditional to modern fuels during the last decade. To obtain a profile of the transition in fuel use over the years, we compare the results of the 1994 study with those obtained in a study conducted for Hyderabad City during the period of 1981–1982 (Alam et al. 1985b).

Overall Patterns of Total Energy Use

The total demand for energy in the household sector is determined by the demand of the various energy-dependent domestic functions such as cooking, water heating, lighting, air cooling, and entertainment. A mix of commercial and traditional energy sources meets this energy demand. The survey reveals that the predominant commercial fuels are electricity and petroleum

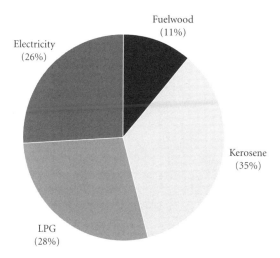

FIGURE 6-1. Fuel Sources for Household Energy Use, Hyderabad, 1994 (total energy use = 25.6 Million KgOE per month)

Source: World Bank 1999.

products in the form of kerosene and LPG. Fuelwood is the main traditional energy source used, while agricultural residue, sawdust, dung, and charcoal are a negligible proportion (less than 1 percent) of the total fuel mix. The main fuels studied in this chapter are therefore fuelwood, kerosene, LPG, and electricity.

The total household energy consumption (including fuelwood) in Hyderabad at the time of the 1994 survey was about 2,500 metric tons of oil equivalent per month. The proportionate share of the different fuels used is illustrated in Figure 6-1. As indicated, the use of fuelwood is rapidly disappearing and now constitutes just over 10 percent of total household energy consumption (this amount is even less when the relative efficiency of fuels is taken into consideration). As will be illustrated later, this small share of the fuel mix is a dramatic change compared with 1982.

The overall patterns of energy use hide rather significant differences in the energy used among income groups. Both kerosene and fuelwood dominate energy use in the lowest 40 to 50 percent of the population by income (see Figure 6-2). The use of fuelwood declines rather quickly as incomes rise and is now used by only a very small percentage of the population. On the other hand, kerosene is now a staple fuel for low-income households in Hyderabad. During the last 15 years, most low-income households using fuelwood have now switched to kerosene. Although these populations use some electricity and LPG, their level of use is rather low compared with more wealthy households. As shown in Figure 6-2, the use of both LPG and electricity are very dependent on the level of income, starting in the middle income ranges. The

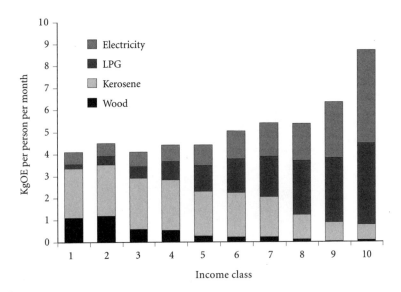

FIGURE 6-2. Energy Use by Income Class (KgOE/Person/Month)

use of LPG and electricity accounts for about 50 percent of total energy use in the middle income ranges, and for more than 90 percent of total energy use in the highest income groups.

The overall patterns of energy use follows a classic pattern found in India and (with slight variation) in other countries as well. The poor are reliant on wood and kerosene, and they use less energy than do more wealthy households. The poor with ration cards do take full advantage of the subsidized kerosene, consuming virtually the equivalent of one month's allotted supply of kerosene (15 liters per household per month). Notwithstanding, and consistent with the results obtained in Chapter 4, the percentage of income poor people spend on energy is higher than that of more wealthy households. Before turning to these broader issues, we examine changing patterns of energy demand in Hyderabad.

Changing Demand for Cooking Fuel

The energy efficiency of cooking has improved significantly as a consequence of the increasing use of kerosene and LPG stoves. Households consumed about the same amount of useful energy for cooking in 1994 as they did in 1982. This is logical, as only so much useful energy is required to cook food. However, the amount of energy used to produce the approximately 9.5 kgoe of useful energy for household cooking was 30 percent less in 1994 compared with 1982 (see Table 6-1). The reason is that the switch from fuelwood to both kerosene and LPG resulted in lower requirements for input energy. An

TABLE 6-1. Average Monthly Household Input and Useful Cooking Energy Consumption by Fuel Type, Hyderabad, 1982 and 1994

Fuel type	Input energy (kgoe/month/household)		Useful energy (kgoe/month/household)	
	1982	*1994*	*1982*	*1994*
Fuelwood	13.3	2.1	2.0	0.3
Kerosene	8.4	11.0	2.9	3.9
LPG	7.0	8.9	4.2	5.3
Total cooking energy	28.8	22.0	9.2	9.5

Sources: Alam et al. 1985b, World Bank 1999.

additional benefit is that harmful fuel-based smoke and cooking fumes also have significantly decreased in urban households in Hyderabad.

The factors that influence the transition from fuelwood are the greater availability of fuels like kerosene and LPG and the relative prices of those fuels vis-à-vis fuelwood. As noted, while the government subsidizes kerosene and LPG prices, market forces determine the cost of fuelwood, especially in urban areas. The increasing scarcity of biomass supplies has resulted in an increasing price for fuelwood. In Hyderabad, the price of fuelwood increased from just above Rs 5 to more than Rs 15 per kgoe for useful energy between 1981 and 1994 (see Figure 6-3). By contrast, both kerosene and LPG increased

Rs. Per KgOE of cooking energy

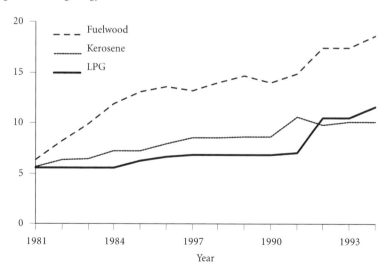

FIGURE 6-3. Useful Energy Price of Cooking Fuels, Hyderabad, 1981–94

Note: Fuel price trends have been adjusted for end use efficiency.

Source: World Bank 1999.

TABLE 6-2. Stove and Connection/Bottle Costs for Cooking Fuels, Hyderabad, 1994

Cost type and fuel	Cost (rupees)
Stove	
Wood	Approx. 50
Kerosene (ordinary)	80–90
Kerosene (pressure)	120–150
Kerosene (wick)	135–150
LPG stove	1,500–3,000
LPG national sector connection charges	
Deposit	900
Regulator	100
Cylinder	108
LPG private sector connection charges	
Deposit and regulator	1,950
Second cylinder	189

Source: World Bank 1999.

from just over Rs 5 to between Rs 9 and 10 per kgoe in the same period. These figures show that the price of fuelwood in Hyderabad has increased more rapidly than the prices of both LPG and kerosene, providing an incentive to fuel switch from fuelwood to kerosene and LPG. On the basis of daily household operating expenses, modern fuels such as kerosene and LPG have become much more affordable than wood for cooking.

In addition to purchasing more modern cooking fuels, consumers also must purchase stoves and in some instances pay for service initiation fees. The costs of such appliances and fees are an important reason why more people had not switched to LPG for cooking by the time of the 1994 study (see Table 6-2). The cost of a wood stove was typically only about Rs 50, and in some instances, consumers could make their own stoves with a few stones or bricks. Kerosene stoves ranged in price from Rs 80 to 130, which is quite affordable. However, the fees for obtaining service for LPG were greater than Rs 1,000 from government suppliers, and about Rs 2,000 for the new private sector suppliers. In addition to these fees, LPG stoves cost in excess of Rs 1,500. The difficulty of qualifying for a service connection combined with the costs of the connection and the stove was a significant barrier for poor people in adopting LPG for cooking.

As a result of the significant differences in the costs of sources of cooking energy, the proportion of households depending exclusively on fuelwood fell drastically, from 13 percent in 1982 to only 1 percent in 1994 (see Figure 6-4). Likewise, the percentage of households employing a mix of wood and kerosene declined from 16 percent to 8 percent. Meanwhile, there was a sig-

TABLE 6-3. Fuel Use Reported by Households for Three Years Prior to 1994 Survey, Hyderabad (Percentage of Consuming Households)

Fuel type	Used more now compared with today	Used less now compared with today	Used same now compared with today	Did not use before today
Kerosene	23	12	64	1
Fuelwood	23	13	61	—
LPG	18	7	70	5

Source: World Bank 1999.

nificant increase in the number of people using both kerosene and LPG. The percentage of people using only kerosene for cooking increased from 28 to 37 percent. The equivalent growth for LPG was from 21 to 33 percent. These figures succinctly illustrate the transition from the use of fuelwood to kerosene and LPG for cooking.

The rate at which the transition is taking place may not be discerned in studies that cover relatively short periods of time. This is demonstrated by the survey data on the levels of fuel use over the three years prior to finalization of the 1994 survey, as shown in Table 6-3. Based on recall questions in the survey, between 60 and 70 percent of all households using a particular energy source recorded no change in their consumption of kerosene, fuelwood, or LPG. However, people generally perceived that they were using more energy at the end of the survey period compared with three years prior. Such changes

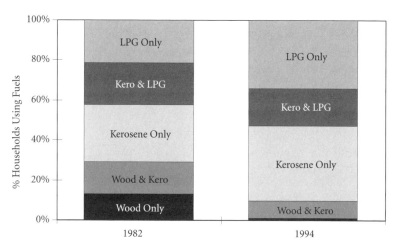

FIGURE 6-4. Changes in Household Choice of Cooking Fuels, Hyderabad, 1982 and 1994

Sources: Alam et al. 1985; World Bank 1999.

in consumption are probably due to some interfuel substitution, changes in family size, or changes in cooking habits due to increasing income.

Near-Universal Electricity Coverage in Hyderabad City

In spite of the rapid growth of metropolitan Hyderabad, the city has achieved near-universal electricity service. The 1994 survey indicates that more than 98 percent of households were grid connected and served by the State Electricity Board. In 1982, the number of households with electricity was officially listed at about 50 percent of the urban population, although actual figures may have been higher because of unregistered connections. Notwithstanding the significant problems with the quality of electricity service, it is evident from the survey that virtually all people in the city benefit from the availability of electricity.

A Shift from Coal to Fuelwood in the Commercial Sector

The decline in fuelwood use by households in urban Hyderabad documented in the 1994 study was not duplicated in the commercial sector. Commercial establishments have been reducing their use of coal and increasing their use of fuelwood as a source of process heat energy (see Table 6-4). In informal interviews with commercial entities, the reasons given for the reduction in the use of coal were that the quality of coal had declined significantly during the last 15 years, and that fuelwood leaves behind much less ash than coal. This shift was confirmed in interviews with wood wholesalers and retailers, who said that their customer base had shifted from households to commercial establishments such as restaurants, bakeries, and ceremonial halls.

The estimated demand for fuelwood is documented in both the household surveys and in a survey of wood retailers in metropolitan Hyderabad. Once again, comparable data were available for both 1982 and 1994 to allow for comparisons of fuelwood demand over time. The figures indicate a marked decline of more than 60,000 metric tons in overall use of wood for energy in urban households between 1982 and 1994. By contrast, both the commercial and the social and religious sectors each increased their use of wood by about 30,000 metric tons during the same period. These overall results indicate that, even though the city of Hyderabad grew by more than 3 million people between the two study periods, the level of wood consumption in the city was stable.

TABLE 6-4. Changes in Sectoral Demand for Fuelwood, Hyderabad, 1982 and 1994

Demand Sector	Wood consumption (metric tons/year)		Compound growth rate, 1982–1994	
	1982	1994	% total change	% annual change
Household	154,031	92,499	51	4
Commercial establishments[a]	13,700	43,015	114	10
Social/religious	10,000	34,368	123	10
Total	177,731	169,882	5	0.3

[a] Includes hostels.

Note: The rates of change are compound growth rates.

Sources: Alam 1985b; World Bank 1999.

Energy Use and Energy Expenditure

Energy is a very important component of consumer spending in Hyderabad, especially for the urban poor. The total overall household spending on energy per month in Hyderabad in 1994 was Rs 212, or about 8 percent of total income. As indicated in Figure 6-5, electricity comprised the largest share of income spent on energy for the average household in Hyderabad, followed by LPG and kerosene. Electricity was used for lighting and for appliances such as fans, televisions, radios, and stereos; it was used rarely for cooking. By con-

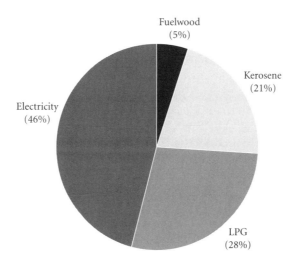

FIGURE 6-5. Energy Expenses for Average Household, by Type of Energy, Hyderabad, 1994

Note: Energy expenses absorb 8 percent of income for the average household in Hyderabad.

Source: World Bank 1999.

TABLE 6-5. Monthly Household Expenditures on Energy by Income Class, Hyderabad, 1994

| Income decile | Rupees per household per month | | | | | % of |
(Rs/capita/month)	Wood	Kerosene	LPG	Electricity	Total	income
<185	23.9	69.7	12.7	48.9	162.6	15.4
185–250	25.6	64.5	20.3	54.0	170.6	11.2
250–300	11.4	60.1	25.9	53.9	157.0	8.2
300–375	13.2	57.9	41.1	57.8	182.5	7.8
375–498	4.4	47.0	57.2	72.6	190.7	7.2
498–583	3.6	38.9	65.8	85.2	195.4	6.8
583–725	4.1	38.1	76.9	114.0	235.5	6.2
725–990	2.0	22.1	92.1	118.7	234.8	5.6
990–1,480	0.1	12.7	95.3	125.2	235.6	4.6
>1,480	0.6	9.0	102.9	222.2	338.4	3.7
Average	9.4	42.3	58.3	96.3	212.0	7.7

Source: World Bank 1999.

trast, the other main fuels—LPG and kerosene—were used primarily for cooking. Thus, the picture of energy expenditures in Hyderabad again confirms that wood was no longer an important household fuel. Wood was virtually squeezed out of household budgets, except for the poorest households. By contrast, the use of electricity grew as an energy expense, even though it was not utilized for cooking.

The overall pattern is that lower-income groups in 1994 still used less energy than higher-income households, but energy was a very significant part of lower-income household budgets. As detailed in Table 6-5 and Figure 6-6, the single largest expenditure for energy spending by the poor was for kerosene, used mainly for cooking. The amount of kerosene available through the ration card program for the poor was 15 liters per month, which resulted in a cost per household of about Rs 45 per month. Poor households, which typically spent more than Rs 50 per month on kerosene, were therefore supplementing their subsidized supplies with purchases on the open market at nonsubsidized prices. Once again, the figures indicate that fuelwood was no longer a significant component of energy spending, and even the poorest households used a significant amount of kerosene for cooking. It is also clear that LPG use was highly dependent on income. The figures for expenditures for LPG demonstrate steady increases, from an average of Rs 12 per month for the poorest households, to more than Rs 100 per month in the highest-income households.

Electricity consumption comprises another significant energy expense for the urban poor. The urban poor spent on average just less than Rs 50 per

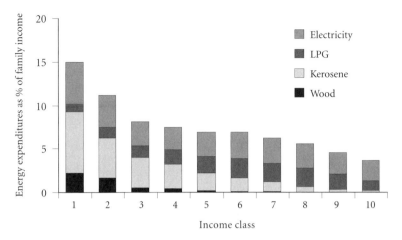

FIGURE 6-6. Household Energy Expenditures by Income Class, Hyderabad, 1994

Note: Income classes are in rupees per household per month and are as follows:
1 = <185, 2 = 185–250, 3 = 251–300, 4 = 301–375, 5 = 376–498, 6 = 499–583, 7 = 584–725, 8 = 726–990, 9 = 991–1480, 10 = >1480.

Source: World Bank 1999.

month on electricity in 1994. The pattern of electricity expenditures is interesting, because, as demonstrated in Table 6-5, the amount of money spent on electricity was relatively stable through the four lowest income classes. It then began to rise with income. This result suggests that the poorest households were using electricity only for very basic needs, such as lighting.

As Table 6-5 demonstrates, the poor spent a high proportion of their income on energy. In the poorest income groups, households paid as much as 15 percent of their income on energy, mainly for kerosene for cooking and electricity for lighting (also see Figure 6-6). The poorest households paid close to 5 percent of their income on the very little electricity they use for lighting. Most other urban households in Hyderabad paid between 2.5 to 3 percent of their income on electricity, and they used much more electricity overall. Focus group interviews of poor households revealed that, due to income constraints, people were already conservative in their use of energy. They tended to turn off lights when they exited rooms and generally conserved electricity use to reduce their bills. The poor also opposed any changes that would result in rising electricity prices.

Expenditures on kerosene constituted about 7 percent of budget expenses for the lowest-income households. The percentage was much lower for the highest income groups (Figure 6-6), who used LPG for most of their cooking and therefore did not require the full amount of kerosene available through the ration program. Conversely, most people in the lowest income groups took full advantage of the government kerosene subsidy program and had to purchase additional kerosene on the open market at world market prices to meet their total fuel requirements for cooking. As might be expected, the focus group interviews revealed that most poor people were dissatisfied with the erratic and inconsistent supply of kerosene coming from the public distribution system, but they were opposed to privatization that would raise prices significantly.

Middle-class households spent between 6 and 8 percent of their income on energy in 1994, and they tended to be concerned with both energy prices and service reliability. One focus group of working women understood that privatization of electricity services would both improve service and mean higher energy prices. The women were mixed in their opinions concerning whether the convenience of better service was worth the higher prices they would have to pay. Some in the group felt that the price of electricity was already high, and that further increases in prices would impose too much hardship. Others agreed with energy privatization in principle, but they were worried about how private companies would deal with safety issues. However, all agreed that privatization would increase the efficiency of the distribution system.

The main energy expenditures of the highest income groups were for electricity and LPG cooking fuel. The budget expenditures on energy for these groups were in the range of 3 to 5 percent. In focus group interviews, higher-income households were more worried about problems related to the supply of LPG and electricity rather than the prices or subsidies. Distributors' delays in replacing empty cylinders was one supply difficulty often cited. People in higher-income groups also strongly complained about the many voltage fluctuations and power cuts. In general, they strongly favored privatization of electricity and LPG distribution. They were aware of the better service in cities such as Bombay and Calcutta, which had private distribution companies. They felt that private industrial groups should be entrusted with the responsibility of ensuring better service.

Energy Subsidies

Government policies have been effective in keeping the price of kerosene low. The average price of kerosene paid by households in Hyderabad in 1994 was about Rs 3.4 per liter compared to a world market price of Rs 5.5 (see Table 6-6). However, the policy to make kerosene affordable actually resulted in poor people paying slightly higher prices for kerosene than did more wealthy families. The reason is straightforward. The kerosene ration of 15 liters per month was not enough for cooking—for which the average family needed about 20 liters. As a consequence, poor families without the ability to afford the initial costs of LPG turned to the informal markets, where fuels are rela-

TABLE 6-6. Comparative Prices Paid for Household Energy by Income Class, Hyderabad, 1994

Income decile (Rs/capita/month)	Price of energy reported by consumers			
	Wood (Rs/kg)	Kerosene (Rs/ltr)	LPG (Rs/14.2 kg)	Electricity (Rs/kWh)
<185	1.14	3.51	107	1.06
185–250	1.15	3.64	108	1.05
250–300	1.26	3.53	107	1.07
300–375	1.26	3.39	107	1.07
375–498	1.80	3.27	108	1.02
498–583	1.50	3.29	107	1.06
583–725	1.35	3.33	107	1.02
725–990	1.72	3.34	107	1.03
990–1,480	1.44	3.30	107	1.04
>1,480	0.82	3.26	107	1.04
Market or reference price	1.23	5.50	175	1.50

Source: World Bank 1999.

tively expensive, to purchase kerosene or wood. Kerosene was available in the informal market for about Rs 5 to 7 per liter. The combination of fuel purchases from ration shops and from informal markets effectively raised the price for poor people compared with households in the middle-income groups, which were able to use a combination of subsidized kerosene and subsidized LPG.

In contrast to kerosene, the prices of both electricity and LPG were nearly the same across income groups. In 1994, the official price of LPG was Rs 107 per bottle, and this is the price consumers were paying because few people were buying LPG from private retailers. The results of the survey regarding electricity usage were somewhat unexpected, because the electricity company charged consumers increasing rates for increasing levels of consumption. For example, the initial block at a use level of 50 kWh cost Rs 0.70, while the highest-use block—usage above 500 kWh—cost Rs 1.45. However, as demonstrated in Table 6-7, the price per kilowatt hour actually paid varied very little across income groups. This is probably due to a monthly service charge for the meter, which raised the effective price of electricity for poor households. The service charge was regressive for poor urban consumers, because it raised the overall price of electric power for those in the lowest electricity-use categories, who were mostly poor people.

The effect of these relatively constant prices on the subsidy levels received by different income groups is that the poor received fewer overall subsidies, because they used less energy than households in higher-income groups. These findings are confirmed when comparing the total subsidies per house-

TABLE 6-7. Per Household Energy Subsidies by Income Class, Hyderabad, 1994

Income decile (Rs/capita/month)	Rupees per household per month					Total Energy Expenses
	Wood	Kerosene	LPG	Electricity	Total	
<185	0	36	7	21	64	162.6
185–250	0	32	12	22	64	170.6
250–300	0	31	15	24	70	157.0
300–375	0	34	24	26	83	182.5
375–498	0	29	33	31	93	190.7
498–583	0	24	38	36	96	195.4
583–725	0	22	45	40	107	235.5
725–990	0	13	54	46	113	234.8
990–1,480	0	8	56	53	115	235.6
>1,480	0	6	61	87	153	338.4

Source: World Bank 1999.

hold in different income classes (see Table 6-7). The poorest households received on average an aggregate subsidy of Rs 64 per month, most of which was derived through the kerosene purchased at ration shops. The main subsidies received by the highest-income households were derived from their relatively heavy use of LPG and electricity; they received total subsidies of Rs 153 per month. This is well over twice the level of subsidies received by poor households in the lowest income groups. If one assumes Rs 1.5 per kWh as a reasonable price to pay for electricity (based on average costs for the electricity company), then the data from Table 6-8 demonstrate that the highest income groups were receiving close to Rs 90 of subsidy per month for electricity, which is an amount that they could have easily afforded to pay.

Although there are some problems with kerosene rationing, Table 6-8 indicates it to be the most effective policy intervention in terms of reaching poor households. The two poorest income groups received a subsidy of close to Rs 7 million per month through this program in 1994, while the highest 20 percent of households, which did not use kerosene very much, received only slightly more that Rs 1 million per month as a class. In fact, it was a well-known informal practice for higher-income households to give their share of kerosene rations to their lower-income servants. However, the highest two income groups are well compensated through other subsidies, as they receive more than Rs 22 million per month in subsidies for electricity and LPG combined.

These findings indicate that energy subsidies were not well targeted in Hyderabad. Both the electricity and the LPG subsidies substantially benefited the more well-off households, with poor households deriving little benefit.[10]

TABLE 6-8. Aggregate Household Energy Subsidies by Income Class, Hyderabad, 1994

Income decile (Rs/capita/month)	Million rupees per month				
	Wood	*Kerosene*	*LPG*	*Electricity*	*Total*
<185	0	3.1	0.6	1.8	5.5
185–250	0	3.6	1.4	2.4	7.4
250–300	0	2.1	1.0	1.6	4.7
300–375	0	2.9	2.0	2.2	7.2
375–498	0	2.4	2.7	2.5	7.6
498–583	0	2.1	3.2	3.0	8.1
583–725	0	2.0	4.0	3.6	9.6
725–990	0	0.9	4.0	3.4	8.3
990–1,480	0	0.8	5.3	5.0	11.1
>1,480	0	0.5	5.2	7.3	13.0
Total	0	20.4	29.4	33.1	82.5

Source: World Bank 1999.

Knowledge of Energy Subsidies

People's perceptions of energy pricing and the fairness of energy pricing policies can be very important for the implementation of changes in energy policy. It is evident from Table 6-9 that very few people knew the basis of energy pricing for the fuels they used every day in their homes. This finding should be qualified by the observation that people who did not use a particular fuel responded "do not know" to such questions. Even with this qualification, the numbers are fairly dramatic. More than 70 percent of households did not know whether LPG or electricity was subsidized.

However, many people did know that kerosene was subsidized. The reason for this is that kerosene bought through ration shops could be compared to more expensive kerosene bought on the open market. As a consequence, one-third of the sample and one-half of higher-income consumers realized that kerosene was subsidized. But while these higher-income households were receiving the greatest amount of energy subsidies per family; remarkably, one-fifth of this same group thought that they were being taxed for this energy.

Clearly, the energy companies supplying fuels for urban households did not communicate the basis for energy prices and the level of energy subsidies. In such an environment, it would not be surprising for consumers to react negatively to any prospect of reducing subsidies and increasing prices for the energy they use in their households on a daily basis.

TABLE 6-9. Consumers' Knowledge of Energy Subsidies, Hyderabad, 1994 (percentage of sample population)

| | "Is fuel taxed or subsidized?" | | | |
Fuel type	Taxed	Neither taxed nor subsidized	Subsidized	Do not know
Kerosene bought in ration shops:				
Average for sample	7	3	35	55
Poorest 10% of population	5	4	25	66
Richest 10% of population	6	1	49	44
LPG purchased through distributors:				
Average for sample	9	3	16	72
Poorest 10% of population	2	0	3	95
Richest 10% of population	20	5	35	41
Electricity from state electricity board:				
Average for sample	13	4	9	74
Poorest 10% of population	.9	0	190	
Richest 10% of population	18	7	19	56

Source: World Bank 1999.

Biomass Demand in Hyderabad and Catchment Area

For the city of Hyderabad, there have been a series of studies on fuelwood markets. This provides the unique opportunity to examine the changes in biomass use in the city over time. Most other work on interfuel substitution only involves measurements of fuelwood supply and demand for only one point in time. Thus, these historic studies provide a unique opportunity to examine how interfuel substitution affects the demand for fuelwood and, by implication, deforestation in the surrounding countryside.

Fuelwood Demand in Hyderabad

The substitution of kerosene and LPG for fuelwood in the household sector actually resulted in declining demand for fuelwood in Hyderabad for the household sector between 1982 and 1994. In absolute terms, household demand declined by 61,000 metric tons, at the rate of a little over 5,000 metric tons per year (see Table 6-10 and Figure 6-7). While the population of Hyderabad increased by more than 90 percent between 1982 and 1994 to five million, household demand for fuelwood decreased by 50 percent, and demand for charcoal decreased by nearly 60 percent. However, the demand for fuelwood in the nonhousehold sector increased, somewhat in line with the population increase, but not sufficiently enough to counter the decrease

TABLE 6-10.   Estimates of Sectoral Demand for Fuelwood, Hyderabad, 1982 and 1994

Demand sector	1982		1994		Change 1982–1994	
	Metric tons/year	*% of total*	*Metric tons/ year*	*% of total*	*Metric tons/year*	*% of total*
Household	154,031	87	92,499	55	61,532	40
Commercial establishments	13,700	8	43,015	25	29,315	214
Social/religious	10,000	5	34,368	20	24,368	244
Total	177,731	100	169,882	100	7,849	4

Sources: Alam 1985b; World Bank 1999.

in household demand. The demand in the commercial and social/religious sectors rose steeply, at the rate of 2,400 and 2,000 metric tons per year, respectively. In 1982, the household sector provided 87 percent of the total demand. By 1994, its share dropped to 55 percent. By contrast, the respective shares of the commercial and social/religious sectors rose sharply from 8 and 5 percent in 1982, to 25 and 20 percent—a three-fold increase during the short period of 12 years. The aggregate effect is that the total demand for fuelwood decreased by 23 percent, and that for charcoal by 7 percent, with a combined decline for fuelwood and charcoal of 11 percent.[11]

While the urban demand for fuelwood decreased between 1982 and 1994, the demand for other wood products, particularly poles, saw logs, and sawn wood increased. This increase more than offset the decrease in demand for fuelwood, which means there was an overall increase in urban demand in Hyderabad for wood and wood products.

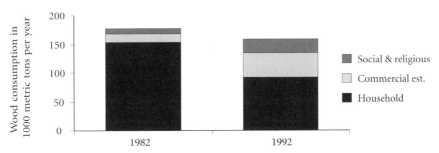

FIGURE 6-7.   Demand for Fuelwood by Sector, Hyderabad, 1982 and 1994
Sources: Alam et al. 1985; World Bank 1999.

Combined Rural and Urban Demand for Fuelwood

Of course, demand for wood does not only emanate from urban areas. People in rural areas use wood in large quantities. Rural populations in the areas surrounding Hyderabad that use wood, straw, and dung for cooking increased by nearly 30 percent between 1982 and 1994, to an estimated 5.9 million people. While much of the rural wood demand, especially fuelwood and poles, is regarded as noncommercial (branches, twigs, and tree tops) compared with urban commercial wood products, all of the former must be taken into consideration when analyzing wood removals from the Hyderabad catchment area. Table 6-11 demonstrates the difference in demand for wood, in terms of roundwood equivalents for the urban and rural populations living within a 100-km radius of Hyderabad.

At the rate of urbanization indicated by the Hyderabad surveys, metropolitan Hyderabad will double in 15 years to a population of about 10 million people. This will lead to a doubling of the housing stock, assuming no change in the average family size. Thus, while it is anticipated that fuelwood will continue to become a marginal household fuel in urban areas, this fall in demand may be more than offset by the increase in fuelwood demand in the nonhousehold sectors. To this should be added the predicted increase in demand for wood in construction, furniture and joinery, and other wood-using industries. There will thus be a moderate but growing demand for all wood products in Hyderabad over the next 15 years.

Geographic Distribution of Fuelwood Supply

The government of Andhra Pradesh has prohibited the supply of wood from government forests. Supplies are now being received mainly from private forests, waste lands, and isolated trees on farmlands. The total annual supply of wood (including both logs and fuelwood) to Hyderabad is 288,000 metric tons, including 172,800 metric tons of fuelwood. In 1982, Hyderabad consumed 177,731 metric tons of fuelwood. There has thus been a small drop in the demand for fuelwood, primarily due to a dramatic decline in the demand from the household sector, from 154,031 metric tons in 1982 to 92,498 metric tons in 1994. The total value of trade at Hyderabad wood auction centers increased from Rs 35.7 million in 1982 to Rs 250 million in 1994, a value increase of 600 percent in little over a decade.

The sources of wood supply are widely distributed. The distances from which fuelwood is brought to the city vary between 63 and 136 km. There was hardly any change from 1982 to 1994 in the average distance and source of supply of wood to Hyderabad City, although the total supplies increased significantly. The bulk of the wood to the auction centers was brought mainly from the adjoining districts. Two opposite forces appear to have exerted pres-

TABLE 6-11. Estimated Rural and Urban Demand for Wood, Hyderabad Catchment Area, 1982 and 1994 (million metric tons, roundwood equivalent)

	1982				1994				Difference (1982–1994)	
	Rural	Hyderabad	All urban	Total	Rural	Hyderabad	All urban	Total	Urban	Total
Population (millions)	4.63	2.59	3.09	7.72	5.90	4.97	5.75	11.65	2.66	3.93
Wood type										
Fuelwood	1.19	0.24	0.29	1.48	1.51	0.22	0.25	1.76	0.04	0.28
Charcoal	0.01	0.01[a]	0.01[a]	0.02	0.01	0.01[a]	0.01a	0.02	0.00	0.00
Poles	0.14	0.02	0.02	0.16	0.18	0.04	0.05	0.23	0.03	0.07
Saw logs	0.07	0.05[b]	0.06[b]	0.13	0.09	0.08[b]	0.09[b]	0.18	0.03	0.05
Total wood[e]	1.41[d]	0.32[c]	0.38[c]	1.79	1.79[d]	0.35[c]	0.40[c]	2.19	0.02	0.40

Notes: Hyderabad catchment area refers to the city and its surrounding periurban areas. Catchment area encompasses all urban and rural populations living within a 100-km radius of Hyderabad. Also, it is assumed that some charcoal, saw logs, and sawn wood, plus all panel and paper products come from outside the 100-km radius of Hyderabad, estimated as follows:

[a] Total charcoal wood, 0.12 million metric tons (mmt), 85 percent from outside Andhra Pradesh (Alam et al. 1985b). Of the charcoal from Andhra Pradesh, only about half was from within the 100-km radius.

[b] For 1981, total saw logs, 0.11 mmt. This excludes 0.04 mmt of off-cuts already counted in fuelwood total. For 1994, total saw logs, 0.20 mmt. This excludes 0.07 mmt of off-cuts already counted in the fuelwood total.

[c] For 1981, total roundwood demand, 0.54 mmt. This excludes 0.03 mmt of roundwood used for panel and paper production, consumed in Hyderabad and other urban areas. For 1994, total roundwood demand, 0.62 mmt. This excludes 0.06 mmt of roundwood used for panel and paper production, consumed in Hyderabad and other urban areas.

[d] For 1981, this excludes 0.02 mmt of roundwood used for panel and paper production consumed within the rural 100-km radius catchment area. For 1994, this excludes 0.02 mmt of roundwood used for panel and paper production consumed in the rural 100-km radius catchment area.

[e] For 1981, estimated total demand, in roundwood equivalent terms, for all wood products, including panels and paper, in mmt: Rural 1.43; Urban 0.57; Total 2.00. For 1994, estimated total demand in roundwood equivalent terms for all wood products, including panels and paper, in mmt: Rural 1.81; Urban 0.68; Total 2.49.

Sources: Alam 1985b; World Bank 1999.

sure on the areas of wood supply. The reduction in the forest area and tree cover on the outskirts of the city pushed the area of supply farther away from the city, while rising transport costs forced suppliers to bring logs and fuelwood from the nearest available sources. Notably, bullock carts, which were used prominently to transport chipped wood to the city in 1982, ceased to operate by 1994.

The Hyderabad wood market comes close to the classical competitive market model. There are a large number of sellers and purchasers in the mar-

TABLE 6-12. Estimated Population, Hyderabad and Vicinity, Selected Years, 1894–1994 (millions)

Year rural/urban	1894	1928	1963	1971	1981	1991	1994
50-km radius of Hyderabad							
Rural	0.51	0.56	0.82	0.95	1.15	1.34	1.40
Urban	0.28	0.57	1.40	1.92	2.71	4.43	5.17
Total	0.79	1.13	2.22	2.87	3.86	5.77	6.57
100-km radius of Hyderabad							
Rural	2.05	2.25	3.29	3.84	4.63	5.59	5.90
Urban	0.33	0.68	1.64	2.20	3.09	4.98	5.75
Total	2.38	2.93	4.93	6.04	7.72	10.57	11.65

Notes: The population was estimated based on the populations of the following districts in proportion to their area within the specified radii: Hyderabad, Rangareddi, Medak, Nalgonda, and Mahbub Nagar. The small areas of Warangal and Nizamabad within the 100-km radius were included in Nalgonda and Medak, respectively.

Sources: Alam 1985b; World Bank 1999.

ket. It is also interesting to note that business worth millions of rupees is transacted every day in these markets, and most transactions are on a cash basis. The supply-and-demand basis of the wood market makes it flexible and ensures its smooth functioning.

Changes to Forest and Wooded Areas in the Hyderabad Hinterlands

Over the last 100 years, the population of India has increased four-fold, from just over 225 million to more than 930 million. In another 50 years, it may be the most populous nation on earth. This tremendous increase in population has and will continue to have a significant impact on land use and natural resources, both renewable and nonrenewable. One hundred years ago, 90 percent of the population lived in rural areas; today it is about 60 percent. These statistics are mirrored by the population living in and around Hyderabad. Table 6-12 provides estimates of the rural and urban population within radii of 50 and 100 km of Hyderabad between 1894 and 1994.

While the urban population of India, including that of Hyderabad, has increased nearly 20 times in 100 years, the rural population has increased less than three times. Thus, within a 100-km radius of Hyderabad, the combined rural/urban increase has been nearly five-fold over the 100-year period, whereas within a 50-km radius, it has been more than eight-fold.

The need for more food and wood products to satisfy the requirements of this increase in population has affected the quantity and quality of forest and

TABLE 6-13. Land Use Changes within a 50-kilometer Radius of Hyderabad, Selected Years, 1928–1994 (all figures in square kilometers)

Land use type	1928	1963	1981	1994
Wooded land	2,073	949	807	752
Forest	1,055	530	445	365
Scrub land	1,018	419	362	387
Agricultural land	4,486	5,620	5,518	5,470
Arable	2,855	3,600	3,466	3,098
Fallow	900	1,133	1,290	1,641
Pastoral	731	887	762	731
Barren/rocky land	75	498	474	470
Built-up areas	28	70	135	257
Water bodies	160	135	130	124
Miscellaneous: non-agric.	435	585	793	784
Total	7,857	7,857	7,857	7,857

Notes: As with the estimate for population, the land uses in Hyderabad, Rangareddi, Medak, Nalgonda, and Mahbub Nagar were tabulated and divided in proportion to the areas within the 50-km radius of Hyderabad. The built-up areas were estimated using the 1994 figure based on the urban population at the specific date. Similarly, the estimates of water bodies were made from 1985–1986 and 1994 data. Miscellaneous non-agricultural land is a residual figure and could include all categories of non-agricultural land.

Sources: Alam 1985b; World Bank 1999.

woodland areas within a 50-mile radius of Hyderabad (see Table 6-13). In the 35-year period from 1928 to 1963, the forest area shrank by nearly half and the scrub land decreased by 60 percent, for a combined total loss of more than 112,000 ha of wooded area—more than 3,200 ha per year. Most, if not all, of this land was converted into agricultural use to meet the growing food demands of the increased population of more than 1 million people. In the 31-year period from 1963 to 1994, the loss of wooded areas declined significantly, by nearly 640 ha per year, despite a population increase of more than 4 million people. In addition, from 1963 to 1994, the land under agriculture decreased by an average of 480 ha per year, with arable land decreasing by more than 1,600 ha per year. This latter decrease was offset by an increase in fallow land, but pastoral land decreased by about 500 ha per year.

Despite this decrease in the arable area, food grain production in India increased substantially, due to a doubling of the area under irrigation, improved seed varieties, and better management. These initiatives doubled the unit grain production per hectare. From 1965 to 1980, the irrigated area increased by 50 percent, and from 1980 to 1994, it increased by another 25 percent. On the other hand, the increase in unit rice production was the reverse: 25 percent from 1965 to 1980 and 50 percent from 1980 to 1994. The main cause of the increase can be attributed to improved seed varieties. This "green revolution" and intensification of agriculture is one of the reasons for the decline in the deforestation rate.

While the rate of deforestation has been declining, the composition of the forests and scrub lands appears to be deteriorating, principally through overuse. There is a gradual but perceptible decline of tree cover in wooded formations from dense forest types to open scrub, via open forests and dense scrub.

Conclusions: Changing Energy Patterns in Hyderabad

The quality of energy service for urban residents in Hyderabad has improved over the last 14 years. More people have electricity than ever before. In addition, households are increasingly switching from wood to kerosene and from kerosene to LPG because of the greater convenience and decreased smoke production in cooking with these fuels.

As the markets for LPG have been liberalized and expanded, many households have taken advantage of clean-burning LPG for cooking. As poor consumers also have moved up the energy ladder to kerosene, very few households in Hyderabad are now using wood as a main fuel. This shift indicates that policies to improve energy access have been moving in the right direction. In spite of the improvements in energy supply and use in the city, however, some bottlenecks remain in the distribution system. There is also a substantial lack of knowledge among consumers concerning energy pricing policies and the extent to which some commonly used fuels are subsidized.

A major issue is the effectiveness of energy subsidies in reaching poor urban households in Hyderabad. Although some subsidies do reach poor people, high-income groups are garnering the vast majority of the subsidies, mainly because they can take advantage of the subsidies for LPG and electric power. The poor, who spend a significant proportion of their income on energy, are being reached mainly through kerosene subsidies. The implications are that the existing structure of energy subsidies at the time of this study was inefficient in meeting the objective of assisting the poor. Recently the public electricity company serving Hyderabad has undertaken corrective measures to resolve this problem.

Increasing urbanization has exercised tremendous pressure on the demand for fuelwood. Evidence of this is notably marked in the hinterland of metropolitan Hyderabad, mainly along the transport routes. Historically, extensive forest lands have been degraded into scrub land, or their densities have been substantially reduced by frequent harvesting. In some areas, total deforestation has taken place as a consequence of urban demand. This reduction of biomass adversely affects overall sustainability, which is a cause for concern.

The combination of the following developments has been very beneficial for forest and scrub areas surrounding Hyderabad: the easing of fuelwood demand in urban areas, the increasingly attractive market for farmers to grow trees, government policies of forest conservation, tree regeneration, social forestry, the prohibition of tree harvesting on government forest lands, and the banning of the outward movement of wood from these forests. There has been a substantial drop in the deforestation rate in the hinterlands of Hyderabad, from 28 km^2 per year during 1928–1967 to only 2 km^2 per year during 1963–1987. The deforestation rate calculated through satellite images more recently reveals an even lower rate of forest decimation, at 1.3 km^2 per year. The satellite data do stress the point that demand for fuelwood in metropolitan Hyderabad exercises considerable pressure on fuelwood resources in the hinterlands.

However, people and communities contemplating growing trees are getting mixed signals from government and local authorities. On the one hand, there are extension efforts that encourage communities and individuals to plant trees and to manage wood lots and natural woodland areas. This should continue and be expanded. On the other hand, barriers that have historically shown to discourage people from growing trees (Arnold 1979) continue to constrain private tree management. In theory, permission has to be obtained to thin or fell trees; in practice, this permission has been relaxed for "nonforest" trees such as neem. But a private individual or a community is not free to fell "forest" species, even if they are on private land. This rule is a deterrent to planting of forest tree species, especially indigenous species, a policy that one arm of government is trying to encourage.

This chapter and previous sections of the book have documented the tremendous diversity in urban energy transitions and the complex factors that drive urban energy consumption in different contexts and at different stages. This complex reality raises the issue of what policies are appropriate for cities at different stages of the energy transition to assure efficient and equitable access to energy services, to help alleviate poverty, and to maintain human and ecosystem health. In the next chapter, the insights gleaned from the research findings in this and previous chapters are applied to develop appropriate policies for cities at different stages of economic development.

7
Toward More Effective Urban Energy Policies

Few societies have traversed a straight-line path from traditional fuel consumption to the use of electricity and other modern fuels. Between the endpoints of the transition process, one finds urban households consuming a variety of fuels and taking a myriad of routes leading more or less rapidly and directly toward modern fuel consumption. The complexity and diversity of the transition process, combined with the piecemeal, case approach to its study has given rise to many viewpoints in the literature with respect to the important determinants and permutations. The lack of a comprehensive view of the energy transition has made it more difficult for governments and international institutions to formulate feasible and cost-effective strategies to help urban dwellers use energy efficiently, affordably, and in ways that promote their own well-being and that of the local and global environments.

Our study is based on synthesis and analysis of survey data from 45 cities in 12 countries, as well as intensive analysis of one case study. This comparison makes it possible to understand the relative effect of key factors influencing the urban energy transition and to draw lessons for policy with regard to the periurban environment, human health, and poverty alleviation.

What Factors Drive the Transition?

Although many factors influence how the energy transition will take place in a particular setting, some key determinants pertain across all cases in our study. These common factors are consumer income, energy pricing, and fuel access.

Consumer Income

Study after study has shown that higher-income consumers predominantly use modern fuels, while lower-income consumers use traditional fuels. Beneath this generalization, however, lies extraordinary variation among countries in the actual income levels at which people switch to modern fuels. This variation has led to some differences of opinion in the literature about how much influence income actually plays in determining the fuels people choose, the amounts they consume, and whether broadly applicable income thresholds can be ascertained for when people transition to modern fuels.

Our study demonstrates that people start switching from wood at surprisingly low incomes—between $12 and $30 per household per month. But it is also apparent that where wood is inexpensive and readily available, people continue to use it extensively, even in more well-off households that earn up to $100 per person per month. The variation in observed income thresholds suggests that income "cut points" for fuel substitution cannot be identified precisely. For the crucial middle-income classes (i.e., excluding the poorest and richest consumers, whose behavior is more constrained or more predictable), it is perhaps better to think of income ranges—with variations within the ranges depending heavily on fuel availability and prices.

Additionally, we find modern fuel consumption at higher-than-anticipated levels among poorer households. This can reflect subsidy policies for some fuels; for example, subsidies for kerosene in Indonesia, coal in China, and LPG in some countries. But it also can stem from the distinctive advantages and attractiveness of the consumed energy sources. Even the very-low-income classes in urban areas are likely to adopt electricity if they have access, due to its superior properties for lighting and running small appliances. In general, the use of modern fuels, including electricity and LPG, intensifies at incomes of about $40 to $50 per person per month. This information should suggest to policymakers that urban energy markets are richer and more diverse than previously thought, and it should provide some encouragement to utility companies and distributors to enhance access.

Energy Pricing

Energy prices in developing countries crucially influence the fuels that urban consumers use. At the early stages of the energy transition, the prices of traditional fuels, such as fuelwood, compete mainly with those of the transition fuels, such as charcoal and kerosene. In these circumstances, people use fuelwood mainly when its price is low compared with the prices of the transition fuel substitutes. When fuelwood becomes scarce or expensive, consumers at this stage switch—often to charcoal.

Charcoal has attributes both of a traditional and modern fuels. When it is produced inefficiently and used in large quantities, it requires large amounts of wood and is consumed in a manner essentially similar to the traditional use of wood. On the other hand, charcoal has a higher energy content on a weight basis than does fuelwood, and when it is produced and consumed efficiently and used in limited quantities for special purposes such as grilling, it is essentially a transitional or modern fuel. The main competitors to charcoal in this second case are kerosene and coal, depending on local resource conditions.

In fact, people often switch fluidly from one to another of these three fuels, depending on local availability and relative prices. Thus, in Indonesia people in the middle income classes use kerosene because its is cheap and available almost everywhere. In China, low- and middle-income consumers use low-grade coal because it is cheap compared with other transition fuels. In Haiti low- and middle-income consumers use charcoal because wood is very distant from urban areas, and modern fuels are too expensive due to petroleum taxes. In some cases, people may switch directly from wood to electricity because of distinctive resource conditions; this was the case in Vientiane, Laos, when abundant new hydropower resources and cheap appliances became rapidly available. Thus, in the middle income ranges—and at the middle stages of the energy transition—people tend to make diverse fuel choices based on the price and availability of substitutes.

At the later stages of the energy transition, many households choose to cook with LPG and in fewer instances electricity—the premium cooking fuels in terms of convenience and flexibility, clean combustion, and service quality. The people who use these fuels are those mostly in the higher income classes, for whom barriers to access and fixed costs are typically not constraining, and those middle-class consumers who can overcome usage constraints and fixed costs.

Research shows that a family's decision to adopt LPG is typically based on a careful comparison of its price with that of fuel alternatives. The reason is that LPG carries high fixed costs involved in purchasing a stove and gas bottles, and price and supply security are also a concern. Also, LPG use is limited to cooking; families, once they adopt the fuel, are likely to use very similar amounts of it, regardless of income. By contrast, virtually all people who have access to electricity choose to use it. Unlike LPG, electricity has many uses— lighting, powering fans, cooking, ironing, air-conditioning, and entertaining. The level of electricity use varies with the price of electricity and the income of the family, and many poor and middle-income families that have access to electricity cannot afford to use it exclusively.

At all but the uppermost income levels (at which consumers will almost certainly commit predominantly to electricity and LPG), energy pricing plays a vital role in fuel choice by urban consumers. Consumers are well aware of the market prices of fuels and switch among them to achieve economies and better service. This is a logical finding, in that most urban dwellers spend from 10 to 20 percent of their cash income for the fuels they use for cooking and basic household purposes. Thus, policies that influence what price-conscious lower- and middle-income consumers pay for particular fuels are often crucial in determining the path that the energy transition follows.

Fuel Access

Consumer access to a fuel is determined fundamentally by two related factors: the availability of the fuel in the local market and the up-front costs of the equipment needed to use the fuel. Well-functioning fuel markets allow consumers a choice among competitive fuels. Governments severely limit such choice in many countries, and the energy transition is slowed accordingly. As noted earlier, high taxes on kerosene virtually eliminate this fuel from the choice set in Haiti for all but the wealthiest consumers. Similarly, the LPG distribution companies in India can service only a small part of total demand due to import restrictions. Customers thus have to register on a waiting list for rights to purchase limited supplies. Likewise, only some regions in Zambia have electric utility service, so people living outside those areas have no opportunity to connect to the electricity grid, regardless of their desire or income.

The effect of the up-front capital costs for some modern fuels represents a crucial factor in the pace of the energy transition. In many southern African countries, consumers are required to pay all of the distribution costs to the utility before they can receive an electricity connection. The poor cannot afford to do so, and thus they must spend more on other fuels, such as kerosene, which provides inferior quality lighting. Because of the special stove and the bulk purchases in large bottles that most distributors offer, LPG is a similar case, though on a smaller scale. Carefully considered policy intervention to alleviate or amortize initial costs could make energy markets distinctly more competitive in developing countries, thus facilitating energy choice and enhancing the quality of life for urban consumers.

Transition and the Environment

Our study suggests that periurban deforestation is associated with urban growth. Rural migrants in urban areas probably contribute by continuing their traditional preferences for biomass fuels. However, continued resource damages are likely even in larger cities that have largely transitioned to modern fuels, due to the scale of aggregate demand. Continued use of woodfuel and charcoal is both common and substantial in such large cities as Bangkok and Manila, due to the sizeable demand from lower-income consumers.

The continued use of biomass fuels also maintains the exposure risk of women and children to indoor air pollution generated from cooking fires. Due to the sizeable population of low-income residents in most larger cities, it seems likely that the total population exposed to indoor pollution will not decline significantly until the very end of the urban energy transition. Policies to accelerate substitution towards modern fuels should have the double benefit of reducing indoor air pollution exposure risks while reducing deforestation pressures on periurban lands.

Policymakers also often seek to ameliorate deforestation by strengthening property rights to prevent uncontrolled resource mining. Or they may reform fuelwood pricing by imposing stumpage fees that more accurately reflect scarcity values. Or they may subsidize reforestation. However, these policies may not always be effective. For example, it may be impossible to enforce property rights on wooded lands where population densities are low. Even where property rights are enforceable, reserving, buying, or protecting the land may not be cost effective. However, the economic development of cities should itself stimulate a more rapid transition away from biomass to modern fuels through higher wages and the resulting increase in the labor costs of fuelwood gathering. It should be remembered as well that biogeographic features affect periurban deforestation—for example, areas with suitable growing conditions for trees can sustain much higher levels of woodcutting than areas in which trees take a long time to grow and produce usable wood. Hence, policymakers should move with due caution and cost consciousness when considering direct controls on biomass mining.

In the largest cities, urban inhabitants consume less per capita biomass due to the recessed resource hinterlands, the intensification of agriculture in cleared periurban land, and the development of modern fuel markets. The leveling off of biomass use on a per capita basis observed for larger cities suggests that the "woodfuels gap" theory is likely to overestimate the rates of biomass consumption and periurban deforestation associated with urban growth.

The need for prudent policymaking is underscored by the fact that our present research could not untangle the relative contributions of several causal factors underlying periurban deforestation. The research does suggest that a combination of urban energy demand for food and wood, conversion of forests for crop production, and fuelwood gathering and charcoaling are important causes of periurban deforestation. But these factors are correlated, and additional research will be needed to clarify the individual roles of various deforestation causes.

Implications for Poverty Policy

Those who are formulating urban energy policy need to pay particular attention to the effects of the energy transition on the poor, a group that is particularly vulnerable to changes in energy availability and prices and particularly burdened by the health and inconvenience cost associated with traditional fuels. The figures on energy expenditures by the poor in our study are from energy surveys and are somewhat higher than found in other studies, which have not measured biomass expenditures in enough detail and hence underestimated them or overestimated the degree to which such fuels are "freely gathered" from the environment. The mistaken impression left is that biomass fuel users are not particularly sensitive to energy price changes. Our study suggests the opposite reality. Therefore, more attention should be paid to the impact of energy policies on the poor.

Do the Poor Pay More for Energy Services?

The cash income of the poor is so low that even modest increases in energy prices impose substantial hardship. The poor spend less on energy than do the higher income classes, but the percentage expenditure is typically much greater—the poor spend between 10 and 20 percent of their income, whereas more well-to-do households spend less than 5 percent. The cost of energy services for the poor is also higher than for higher-income customers because cooking with fuelwood and lighting with kerosene are cost inefficient compared with cooking and lighting with modern fuels. Moreover, the poor often buy fuelwood and charcoal in small amounts, and the higher transaction costs of buying in small quantities inflate the price. Once the comparative efficiencies and transaction costs have been taken into consideration, the delivered energy for cooking often is more expensive for the poor. Inconvenience and health costs further add to the economic costs of energy utilization for lower-income households.

The case of lighting is even more dramatic than for cooking. Electricity provides 15 times more light than the same amount of energy contained in kerosene, and this figure can be as much as 200 times higher, depending on the type of kerosene or electric lamp. Consequently, the price of light for the poor is often higher than the price of light for wealthy households. The one exception is in urban areas with near-universal electrification, where the poor use electricity mainly for lighting. In these areas, the poor pay about the same amount as do other classes. It is uncommon, however, to see the poor pay less for lighting. To our knowledge, only in Thailand, with its system of increasing block rates with increasing usage, do lower-income groups pay less for lighting.

What Kinds of Policy Interventions Work?

Although our study shows that the poor pay a higher percentage of their income on energy services in cities with high energy prices, it also shows that the poor pay a lower percentage of their income on fuels in cities with low energy prices. This offers scope for intervention, but the policy implications—for example, regarding subsidizing of fuel prices—are not clear cut.

Subsidies for modern cooking fuels have both direct and indirect effects on the fuel expenditures of low-income consumers. A subsidy on a particular fuel such as kerosene reduces direct cash outlays for kerosene if it is available. If the kerosene is not available, it may still have an indirect effect because the prices of fuelwood and charcoal often are capped by the prices of more modern fuels. A general fuel subsidy can help the poor, as do the kerosene subsidies in Indonesia and the coal subsidies in China. However, such broadly targeted subsidies benefit all income classes. As such, they are an expensive way to provide relief to the lower income classes.

The undesirable side effects of broad subsidies can in theory be avoided by subsidizing only fuel rationed to the targeted income class. However, these policies have generally worked imperfectly, because the rationed fuel can end up on the black market. This market is often exploited by middlemen or wealthier consumers.

For electricity, targeted subsidies have proved more effective. The lifeline rates offered by many power companies in developing countries do assist the poor because they can target the initial block of consumption, for example, the first 50 kWh per month. According to the survey data in this study, most poor families use electricity mainly for low-demand uses such as lighting, and they consume approximately 50 kWh per month. Those whose meters indicate greater consumption are charged for the additional increments at higher block rates.

Lifeline rates are prudent policy, but they work well only if electricity connections are available to the poor at affordable rates. However, where access is restricted by capital costs and trouble involved in obtaining a connection from the power distribution authorities, the poor often share electricity with a household that has a legitimate, metered connection. But by sharing a connection, they end up paying at higher block rates for their power. Lifeline rates should still be limited to blocks of less than 50 kWh per month to keep the subsidy on target, but they should also be accompanied by an assisted connection policy. That strategy, combined with the attractiveness to households of switching from kerosene to electricity as a high-quality source of energy for lighting and small appliances, can create a substantial impetus in the energy transition.

It should also be noted that just as subsidizing modern fuels has an indirect positive effect on the poor, taxing them has an indirect negative effect.

Many countries try to raise government revenues and encourage energy conservation by taxing modern fuels, which are used primarily by the wealthy. However, like subsidies, taxes on modern fuels affect the level of traditional fuel prices by changing the cap price. Consequently, in countries such as Haiti and Mauritania, where petroleum products are taxed, the price of traditional fuels is very high. Because the poor have difficulty collecting fuels in these resource-constrained countries (see Kumar and Hotchkiss 1988), they end up paying modern fuel prices for their biomass fuels. The only situation in which this is not the case is in countries where the price of biomass fuels is less than the price of modern fuels, which indicates an initial transition stage of abundant wood resources. In most cases, however, taxing rich people's fuels has the indirect effect of hurting the poor by raising the price of the substitute biomass energy.

Energy policies can have a tremendous impact on the limited cash budgets of the poor. The most effective interventions are low block rates or lifeline tariffs aimed at poor households that generally use very little electricity, coupled with efforts to promote electricity access for the urban poor. Credits or subsidies to lower the initial costs of stoves or bulk fuel purchases will help expand the fuel choices of the urban poor. Having such choices will not only help the poor gain access to modern fuels, but fuel choices will also put downward pressure on the prices of the traditional fuels that the poor use most often.

The Role of Government

As the preceding discussion emphasized, government policy is influential in determining the fuels that people use during the transition from wood to modern fuels in urban areas of developing countries. Income accounts only partially for the direction and the shape of the energy transition. Government intervention in the form of pricing measures, import restrictions, regulations on fuel distribution and marketing, and forest and tree management are influential in shaping the energy transition. The diversity of the energy transition arises from the complex interplay between household purchasing power, the functioning of local markets, government intervention, and local natural resource conditions.

One conclusion of our study is that appropriate policies for dealing with the problems that occur during the transition from traditional fuels to modern fuels vary with the stage in the transition process (see Table 7-1). Interventions for countries experiencing rapid deforestation should be different than interventions for cities that have already switched to transition fuels and are faced with shortages of modern fuels. Health problems associated with indoor air pollution also can be an issue for cities in which the use of open wood burning is prevalent (Hughes et al. 2001; Smith et al. 2000), but it may not be a problem in cities using mostly LPG or kerosene for cooking.

Where government intervention is warranted, four complementary strategies are candidates: promoting interfuel substitution, pricing energy to reflect external resource costs, helping rural people to manage existing woodlands, and improving energy efficiency for both traditional and modern fuels.[12] Although the mix of these options will vary with the location, the resources, and the stage of transition of any given city, the group of strategies, taken together, can lead to more rational use of resources.

Interfuel Substitution

Interfuel substitution can be an important tool for alleviating health and environmental problems associated with the consumption of traditional fuels. In the early stages of the transition, it is sensible to consider woodfuel substitutes only when woodfuel prices are competitive with modern fuels or when they are rising rapidly. If wood is abundant and relatively inexpensive—as it is near some of the smaller cities in Zambia—little room remains for interfuel substitution by the poor or even the middle class. Once woodfuel and charcoal prices become fairly competitive with kerosene or coal, these other fuels can be considered as alternatives. Policymakers should begin by examining the availability of alternative fuels, along with the affordability of stoves and other appliances.

As a matter of equity, it is important not to impose taxes on modern fuels or to hinder the distribution of modern fuels because of the price effect of such policies on traditional fuels used by the poor. A tax on modern fuels also extends the consumption of biomass fuels for many middle-class consumers beyond the point at which they would normally fuel switch. Taxation puts additional pressure on the wood resources around cities and may lead to health problems from indoor air pollution, depending on the type of stove used for wood burning.

On the other hand, an important policy action would be to eliminate restrictions or bottlenecks from the distribution of transition fuels such as kerosene or coal. Import restrictions on petroleum products, and subsequent rationing with subsidies (as in India), should be avoided. Not only is such a policy manifestly uneconomic, but the poor have problems obtaining ration cards, and the supply restriction raises the price of traditional fuels. Abandonment of targeted subsidies and loosening of restrictions on imports may be needed to clear the bottlenecks or blocks to adoption of transition fuels. Because the fuels are comparatively attractive and efficient, it may be more productive to provide credit to low-income consumers for the purchase of appliances such as stoves. This also may lead to the improved health of urban dwellers.

At the later stages of the energy transition, the main issues involve the availability of modern fuels, modern appliances, and electricity for the poor. In terms of interfuel substitution, it is important at this stage to ensure that basic electricity service is extended to everyone. Recent work in Indonesia indicates that the policy of extending connections to the poor is sound. Adoption of electricity has yielded a substantial improvement in the quality of life for the urban poor. The availability of other premium fuels such as LPG also should be considered a priority. At this stage, credit for the purchase of energy-efficient appliances that use electricity or LPG may be considered.

Price Incentives

Price incentives can be used to comport the cost of household fuel to their economic cost to the nation, or to achieve specific equity objectives. An environmental tax on biomass fuels might be justified in some cases to internalize externalities, but it would worsen consumer equity in energy markets. Subsidies, when considered, should be carefully evaluated for possible unintended side effects (e.g., free riding by the middle class and the rich when subsidies cannot be targeted specifically to the poor). Another issue is that subsidies for household fuels are detrimental to energy conservation, in the sense that the income effect of the subsidy moves in the opposite direction from the substitution effect. Finally, subsidized fuel can wind up being diverted to other markets. In Pakistan in the early 1990s, the government subsidized kerosene to assist the poor, but much of the kerosene was diverted away from households to the transportation sector. In Ecuador, subsidized kerosene wound up on the black market and was exported to a neighboring country.

It has often been stated that energy subsidies cannot be well targeted to the poor. But modest lifeline electricity rates for the poor may be an exception. Lifeline rates for blocks of approximately 50 kWh or even less will cover basic lighting for the poor, especially in cities in the later stages of the energy transition. However, lifeline rates at the 200 to 300 kWh level common in some countries should be avoided, because this helps the middle class and rich more than the poor. In addition, credit may be considered for assisting the poor to pay for the up-front costs that are involved in initiating electricity service. Because the poor gain great benefits from switching from kerosene to electricity for lighting, a lifeline rate and improved electricity access can yield real improvements in their quality of life.

Managing Wood Resources

People living in cities where fuelwood is used extensively generally have good access to wood and poor access to modern fuels. Wood in such cities is relatively available and inexpensive. At this relatively early stage of development,

TABLE 7-1. Stages of the Energy Transition and Possible Policy Intervention Strategies

Stage	Income/ city size	Sector characteristics	Possible intervention strategies			
			Interfuel substitution	*Pricing policies*	*Management of energy supplies*	*Energy efficiency*
Stage 1 Low Wood use Income/person/month City size ('000s)	$8–$38 40–840	• Extensive fuelwood use is often mined from existing woodlands. • Alternative fuels are more expensive than wood, and their markets are not well developed.	• Consider financial and socioeconomic possibilities of kerosene. • Consider credit for kerosene stove purchase.	• Consider resource tax on mined wood. • No taxes on modern fuels.	• Promote the management of trees and forests by rural people. • Explore bottlenecks for distribution of kerosene and other alternative fuels.	• Promote improved fuelwood and/or kerosene stoves, depending on main fuel in use. • Develop improved stoves aimed at reducing indoor air pollution.
Stage 2A Transition Charcoal use Income/person/month City size	$15–$45 12–1,000	• Charcoal often comes from inefficient kilns in distant forests. • Charcoal markets are well developed, and markets for alternative fuels are not well developed.	• Consider financial and socioeconomic possibilities of kerosene. • Consider credit for kerosene stove purchase.	• Consider resource tax on mined fuelwood. • No taxes on modern fuels.	• Promote the management of trees and forests by rural people. • Explore bottlenecks for distribution of kerosene and other alternative fuels.	• Promote improved charcoal and kerosene stoves for efficiency and reducing indoor air pollution. • Promote efficient charcoal kilns.
Stage 2B Transition Subsidized fuel Income/person/month City size	$15–$27 32–7,900	• Kerosene and coal (mainly China) are subsidized to encourage substitution away from wood. Subsidies involve "free rider effect."	• Consider financial and socioeconomic possibilities of wood, LPG, and electricity. • Extend basic electricity service to poor.	• Subsidies for coal or kerosene should be carefully evaluated. • Institute "lifeline rates" for households using less than 40–50 kWh/month.	• Ensure availability of kerosene. • Reduce barriers to use of LPG.	• Promote efficient appliances and lights. • Promote efficient kerosene and/or coal stoves.

Stage 2C Transition Market priced fuel Income/person/month $32–$77 City size 49–8,150	• Countries pursue a market approach and a wide variety of fuels are used, often accompanied by shortages.	• Examine availability of substitute fuels. • Consider loans for purchase of kerosene and LPG equipment. • Extend basic electricity service to poor.	• The existing market policy should be maintained and encouraged. • Institute "lifeline rates" for households using less than 40–50 kWh/month.	• Ensure availability of kerosene. • Reduce barriers to use of LPG.	• Promote efficient appliances and lights.
Stage 3 High Modern fuels Income/person/month $43–$142 City size 36–6,000	• A high percentage of households uses modern fuels. Fast residential electricity growth. Some traditional fuels used.	• Consider principal substitutes that involve backward switching to kerosene or woodfuels.	• No taxes on modern fuels. • Institute "lifeline rates" for households using less than 40–50 kWh/month.	• Reduce barriers to use of LPG. • Assure good electric service and function of modern fuels markets.	• Promote efficient appliances and lights.

clearing for agriculture often produces a glut of low-priced fuelwood. In some cases, such wood is even burned on site to get rid of it. Typically, access to electricity is restricted, as the resources to produce power and distribute electricity are limited. Consequently, only the middle class and wealthier households use electricity.

Although cities in Africa make up the majority of cities in this category in the study, cities from the Philippines and China also fall in this group. Although some cities in Asia have relatively easy access to wood, it is often harvested faster than it is regrown, and such harvesting cannot be kept up for long without environmental problems.

One approach to the problem of overharvesting is to implement a plan for maintaining the productivity of local woodlands or private farms around urban areas. Such a plan must include appropriate incentives for local people to grow or maintain trees (Barnes and Floor 1999; Cernea 1981). An effective approach was taken in Niger, where communities were given control over natural woodlands and their products if they participated in a program to manage the land (Foley et al. 1997). Such management can also involve tree growing by farmers. The private investment of time and money necessary for land management programs often only is possible if farmers or communities can negotiate favorable prices for trees and other wood products.

Once population densities have increased and the demand for wood from urban centers has intensified, the price of wood in urban markets is likely to approach the level of modern, alternative fuels. At this stage deforestation is often well advanced, and land around cities is being managed for agriculture and other uses. In most such cases, land is no longer managed communally but is instead privately farmed. If consumers still demand wood in such situations, private farmers can be encouraged to grow trees. This has happened without intervention in countries such as Madagascar and the Philippines. A study of the island of Cebu (Philippines) indicates that virtually all of the natural forests have been gone for about 100 years. Nonetheless, 30 percent of the people still use wood, supplied to Cebu City by local private farmers who are growing trees and selling wood through a well-organized market (Bensel 1995; for other parts of the Philippines, see Hymen 1983, 1985). A similar situation exists in Haiti, where farmers are growing trees to produce charcoal. In such circumstances, it is important to support the incentive systems that have spawned wood growing to meet fuelwood needs and to ensure that government policies support rather than hinder the process.

Promoting Energy Efficiency

The main policies that encourage energy efficiency in urban areas of developing countries include pricing, efficient stove programs, improved electricity appliances, and more efficient charcoal production. For cities with extensive fuelwood and charcoal use and relatively high energy prices, distribution of fuel-efficient stoves can save urban households scarce cash income and reduce the aggregate wood demand. In addition, the improved stoves can be designed to reduce indoor air pollution by venting cooking smoke outside of the dwelling. If stove programs work, people have more cash income to spend, the health of women and children is improved due to reduced smoke exposure, and the reduction in the demand for wood energy diminishes pressure on the land surrounding urban areas. For areas in which charcoal is the main urban fuel, more efficient charcoal kilns can dramatically reduce the amount of wood that is necessary to produce a given amount of charcoal. If such programs work, the charcoal producers save time and materials in the production of the same amount of charcoal.

At later stages in the energy transition, people in urban areas increase the number of appliances that they own, and consequently the demand for electricity rises significantly. For electricity, household consumption generally increases faster than any other sector in the economy. As the sector grows from about one-tenth of the total electricity demand common in low-income developing countries to one-third of that demand, the conservation of electricity becomes much more important. However, as opposed to improved stoves and kilns, households have fewer incentives to buy efficient appliances. Although energy efficiency often is important for society, efficient lights and appliances may not pay off to consumers for 5 to 10 years. In this case, it is necessary to develop special incentive or standards programs for successful electricity conservation. Such programs can range from publishing energy efficiency ratings on appliance or rebates on the purchase of energy-efficient equipment, a group of practices that is common in many demand-side management programs.

The government interventions discussed in this section vary depending on the conditions in urban areas for specific countries. It is obvious that any necessary intervention must be based on the specific conditions in the country. However, this list of interventions by energy transition stage are indicative of the possibilities that should be considered. In addition, many of the interventions are interrelated. For instance, it would be inconsistent to promote subsidies and to simultaneously expect consumers to adopt energy conservation practices such as improved stoves or more efficient appliances. In addition, no one set of interventions by itself can stand alone in the face of other conditions that prevent it from working. As a consequence, a narrow focus on pricing, conservation, management of existing wood supplies, or interfuel substitution may actually do more harm than good. It is necessary to understand the interrelationship between such approaches and to implement appropriate strategies based on local conditions and needs.

Implications of the Energy Transition for Social and Environmental Policy

The energy transition is an ongoing process that involves movement from the inefficient use of traditional energy to the efficient use of modern fuels for cooking, heating, lighting, and other uses. People in urban areas in Africa will eventually replace their kerosene lamps with electric lights. In Asia, inefficient refrigerators will be replaced with more efficient models. In Latin America, cooking with electricity and LPG will become more common.

Although there are exceptions, most cities will move through the energy transition slowly and in due course. Transitional problems include stress on wood resources around some urban areas, health problems associated with indoor air pollution from open biomass stoves, low standards of energy service, high prices for wood, poor markets for modern fuels, and poorly conceived government policies. It is important to address these problems as they arise. The long-term solution to these problems cannot be forced indiscriminately on urban dwellers regardless of their stage in the energy transition.

The most important recommendation is that people should be given choices. The best way to do that is for policymakers to support open markets, rather than encumbering markets with constraints or access restrictions. Evidence suggests that competitive markets hold down the price of fuels. If fuelwood or charcoal becomes too expensive because of seasonal availability of wood, then people with choices can switch to kerosene. Likewise, people who lack access to electricity must light with kerosene. In turn, the cost of providing access must be evaluated, but having choices means that people have the freedom to choose fuels based on their value to them. Competitive markets keep the prices of individual fuels low and from fluctuating widely, with both efficiency and equity benefits.

The benefits of using energy are provided through the use of some kind of energy appliances or equipment. But sometimes the energy appliances can be expensive. Thus, the importance of choice also extends to the purchase of such equipment. Often the difficulty lower-income people face in obtaining LPG is that equipment is either expensive or not available. Shortages of LPG cylinders are a fact of life in many countries. Small loans to assist in the purchase of such appliances in some cases can act to increase people's access to fuels. Choices are conditioned by both financial as well as physical barriers to entry. A poor person with limited cash income is unlikely to be able purchase a stove and an LPG cylinder and to make a monthly purchase of 15 kg of LPG. Ways to improve choices for people include lessening the burden of such "lumpy" cash expenditures by rolling the costs into loans or energy charges.

The case for improved fuelwood and charcoal stoves in urban areas can be considered as a step before interfuel substitution. In a sense, improved biomass stoves are a new step on the energy ladder. Such stoves are more expensive than traditional stoves because of their design and materials, but in the right circumstances they can save money for households that adopt them, and they can reduce pressure on biomass resources surrounding cities. Moreover, indoor air pollution is now recognized as a major cause of disease in developing countries by international health organizations. Properly designed stoves can eliminate smoke from urban households and can alleviate such health problems. The right circumstances for the promotion and adoption of improved stoves include cities with high biomass energy prices and in which urban people are using some sort of stove already. Although stove programs are not a panacea, they do improve the choices that are available to people living in urban areas—especially poor people.

The poor in urban areas of developing countries face special problems in meeting their basic energy needs. Many poor people have emigrated from rural areas where energy could be gathered at just the cost of their labor. In urban areas, the poor must purchase this same energy and often burn the fuel in the same inefficient stoves they used in rural areas. In countries with high energy prices, the poor pay a significant proportion of their income on energy. Providing the poor with access to various forms of energy, however, may actually benefit them more than extensive subsidies. The example of the far greater efficiency of fluorescent electricity lamps over lighting with kerosene illustrates this point. Access to basic electricity service also allows a poor household to undertake many more activities in the evenings—children can do homework, adults can sew or listen to a radio. Another reason for expanding energy services to the poor is that it gives them a choice to switch from inefficient and smoky stoves to improved models that save energy, reduce fuel costs, and promote better health for women and children in particular. Although the financing of such service is an issue, the compounding value of access to electricity service for society may be much greater than the cost.

Natural resources around urban areas are often abused at various stages of the energy transition, and policies often are necessary for preventing such abuse. The combination of expanding agriculture and increasing fuelwood or charcoal demand in urban areas adds to deforestation pressures in many countries. It is precisely at this stage that land-use systems are changing rapidly, moving from low- to high-intensity use for producing food and fuelwood for urban centers. In regions with very low intensity of land use, trees are simply harvested from existing woodlands, and in areas with high intensity of land use, farmers often grow trees to sell in urban fuelwood markets. In the transition between these two situations, the key to preventing resource abuse is to improve land-use practices by improving the rights and techniques of farmers and communities to manage local resources. One-dimensional approaches to such problems—such as fuelwood plantations established and maintained by forestry departments—are likely to fail because they do not address the primary issue—the intensification of land use (Naronha 1981). Based on local conditions, the solutions to such problems often involve multisectoral approaches, including agriculture, energy, and the environment.

The problems of cities in developing countries are varied and many. Crowded streets, poor water supply, poor sanitary services, and many other difficulties are very important and cannot be ignored. But these problems should not be allowed to completely overshadow problems associated with urban energy. People in urban areas are often faced with costly energy, indoor air pollution, and few or constrained choices for the basic needs of cooking and lighting. Such problems need to be addressed in diverse ways at different stages in the energy transition. The policies recommended in this book are designed to minimize transitional problems in the evolution of urban energy markets and to improve the quality of urban life for people living in developing countries.

Annex:
Methods and Data

This annex explains the methods and data used in our analysis. The first section describes the data and variables used in the analysis of urban energy consumption. The section following considers the methodology used to construct the variables on forestation patterns and development around urban areas, and it identifies the information sources used to assemble the dataset.

Urban Energy Consumption: Sample and Data

A series of ESMAP household energy surveys provided the primary data used in the analysis. Without the valuable work completed in compiling such surveys, this project would not have been possible. ESMAP and other surveys provide information on household energy use at the level of the household and by different income classes.

The data upon which the analysis is based are from urban household energy surveys in countries in Asia, Africa, Latin America and the Caribbean, and the Middle East. All of the surveys are based on stratified random samples of urban populations. Because there are both small and large cities in the study, the sample size for each urban location varies significantly. However, we have insured that an adequate number of households were sampled by deleting cities for which we questioned whether the number of households adequately represented the urban area. In total, the dataset contains more than 25,000 households in 12 countries, for 45 urban areas.

The compilation and standardization of the dataset took one full year. Because the survey instruments for each country were different and the surveys were completed during different years, the standardization process was given a top priority. All surveys were required to have similar sets of standard variables that had been collected in similar ways. All monetary figures were adjusted for inflation to the year 1988. In order to standardize the pricing information, kerosene was selected as the denominator from which to calculate price ratios, thus eliminating exchange rate problems.

Two standard datasets were constructed for the analysis. The first one involves household-level data. All of the results for this type of analysis are based on the analysis of the variables described below for each urban area. Predictably, the price comparisons from this analysis were disappointing because of the small variation in price between cities. The second dataset involves the aggregation of the household level to income class for each city. The cross-urban analysis is based on data aggregated to income class. This was done for computational convenience and because of the belief that the dataset should represent significant variation between cities. The price comparisons for this analysis were very robust.

Variables

The following variables have been extracted from the databases for analysis and are important for the research. The variables for the cross-urban analysis involve average levels or shares of people in various income classes. For the household analysis in urban areas, the measures are either the actual levels of energy use or the percentage of households using a particular type of energy. Income in the various household surveys is measured as total household expenditures, cash income, or income class, depending on the country. For each urban area, income is adjusted for inflation to 1988 levels. Education is measured as the average number of years of education obtained by the household head per income class for the cross-urban analysis and level of education obtained by the household head for the household-level analysis. The percentage of households using a fuel is measured as the share of people who use fuelwood, charcoal, kerosene, LPG, or electricity for either cooking or lighting. Fuel use is measured as the total energy used in the household. The price information involves the actual prices faced by individual households. The surveys contain information on expenditures for different types of fuels. As a consequence, the percentage of total income spent on energy is included in some of the analysis.

Periurban Biomass Variable Construction

Standing Biomass. The biomass estimation methodology involved the following steps:

1. Vegetation maps were located from a variety of published sources that displayed the spatial distribution of different vegetation classes in the region surrounding the cities (see below for a country-by-country description of source material). Maps ranged in scale from 1:8 million to 1:20 million. The distribution outline of different vegetative classification was digitized using an ATLAS*DRAW/ATLAS*Graphics software package—a quasi-geographic information system package—and plotted, producing detailed computerized mappings of the vegetation patterns around each city in the sample.
2. The ATLAS*DRAW/ATLAS*Graphics package allowed the calculation of the areal extent (measured in hectares) of each vegetation class in five concentric zones around each city. The periurban zones were 0–25 km, 26–50 km, 51–100 km, 101–200 km, and 201–300 km.
3. Vegetation-class-specific "biomass conversion factors," which give above-ground biomass in m^3/ha, were applied to convert the areal calculations into standing biomass estimates associated with each vegetation class in each zone.
4. In the final step, biomass estimates for each vegetation class were aggregated to yield a total figure for standing biomass in each zone. These estimates were then divided by the land area in each zone, yielding averaged, zonal biomass density values.

Note that vegetation maps were generally available, but for different years. Consequently, forest stock estimates represent a historical cross-section across the cities in the sample rather than the status of forest resources today. The vegetation maps reveal the extensive effect of deforestation—the conversion of high-density vegetation categories into lower vegetation types.

The deforestation process also has an intensive dimension—the degradation of a given vegetation type. To account for this feature of the process, biomass conversion factors were adjusted to reflect stand degradation associated with deforestation, consistent with the level observed for the map year. World Bank forestry experts checked these adjustments to assure their validity and consistency across the sample.

Biomass conversion factors in the literature are often for stemwood (wood from the trunk of a tree, rather than from branches, stump, or roots). Total aboveground biomass includes tree crown and understory vegetation as well as tree bole. Methodology described by Brown and Lugo (1984), Openshaw (1986), and Ryan (1992) was used to adjust biomass conversion factors upward to reflect the biomass embodied in crown and understory. Again, technical assistance from World Bank forestry experts was used to assure that this process was conducted in a reliable and conceptually consistent manner across the sample.

Roads. We used primary road networks as a proxy for transportation infrastructure. The total road distance in each zone was calculated and then divided by the zonal land area. This "road density" variable is expressed in units of km/106ha. Road maps were from the *Times Atlas of the World*. Maps of the same scale (1:5 million) were used to avoid bias potentially caused by different scales. Bolivia was on a scale of 1:5.1 million. Maps of Cape Verde and the African continent were of the Lambert Azimuthal Equal-Area projection, Asian countries were of the Mercator projection, and Haiti was mapped in the Conic projection.

Topography. Topography was evaluated in terms of elevation and slope. Calculations of the average elevation and standard deviation are useful in the more proximate periurban zones but lose meaning in the outer rings because the standard deviation is quite high for virtually all cities in the outer rings. Changes in elevation—that is, slope—appeared to be a more relevant measure. A qualitative-scale variable was used to describe changes in elevation in each ring.

All source maps were on a scale of 1:5 million or 1:5.5 million. The scale variable was selected based on three slope criteria:

Scale variable	Variable description
0	Flat plains with relatively small changes in topography, generally less than 500 m over a horizontal distance of 25 km (roughly 2% slope)
1	Areas with moderate slope, having gently rolling hills; approximately 1,000–2,000 m change in elevation over 25–50 km horizontal distance (2–8% slope)
2	Sizable areas with steep slopes, having changes in elevation greater than 2,000 m over 25 km (greater than 8% slope)

Precipitation. Precipitation was evaluated for every zone around each city. Precipitation values are given in terms of millimeters per year and reflect the mean annual precipitation observed in each zone. Data were collected from maps depicting weather station readings spanning 20 to 85 years. Twelve random observations were taken in each ring and used to calculate the average and standard deviation. The scale of these maps varied widely, ranging from 1:1 million to 1:19 million. Mapped precipitation information was unavailable for the Philippines, and therefore a value was determined for each city overall rather than on a per-zone basis.

Biomass Sources by Country

Bolivia

Vegetation Map. *Source:* von Borries (1986). "Mapa Forestal de Bolivia" in *Programa de Regionalizacion Energetica de Bolivia*. La Paz: Organization of American States. Published in 1986 but produced in 1977. Scale: 1:6 million. Projection: Unknown. The *Times Atlas of the World* was used to amend portions of the northwest border that were missing on the von Borries map.

Standing Stock Density Derivations. The von Borries document provided information on mean annual increment (MAI) and net primary productivity (NPP), but not standing biomass density.[13] The FAO report was therefore referenced for biomass density figures, with the assumption that all of the Bolivian forests fell in the Production category. FAO figures were increased by a factor of 2.1 to account for crown and understory, following a standardized procedure to convert wood bole into aboveground biomass. Several vegetation categories appearing on the map produced by von Borries did not have corresponding categories in the FAO report. For these forest types, estimates were made based on relative values for NPP and MAI found in von Borries' work.

von Borries' classification	FAO classification	m^3/ha
Humid Alluvial Forest	Bosque Densos de Tierras Bajas	242
Humid Alluvial Floodplain Forest	NA	182
Humid Montane Forest	Bosques Andinos/Subandinos	126
Dry Alluvial Forest	Chiquitania	105
Dry Montane Forest	NA	105
Transition Forest	NA	45
Savannah	NA	15

Note: NA = not applicable.

Humid Alluvial Floodplain Forest was calculated as 75 percent of the Humid Alluvial Forest density; Dry Montane Forest as 83 percent of the Humid Forest. Savannah was derived from FAO values and adjusted based on von Borries' increment figures and Savannah values in other regions (Openshaw 1986). Plantations were not depicted on this map; where plantations exist, they are presumably categorized with similar vegetative formations.

Botswana

Vegetation Map. *Source:* ETC-Foundation: Biomass Assessment of South African Development Coordination Conference (SADCC) States. p.65. Date: 1987 (work completed 1985–1987). Scale: 1:5 million. Projection: Unknown.

Standing Stock Density Derivations. Standing biomass density estimates were taken completely from the ETC report on the SADCC member states. These values include all aboveground biomass and do not make reductions for merchantable timber, and they therefore do not need to be adjusted for crown and understory. The ETC values of metric tons per hectare were converted into cubic meters per hectare at a rate of 1.4 cubic meters per hectare per 1 metric ton per hectare. Vegetation types were aggregated, and biomass density values were calculated as an aerial weighted average as follows:

Aggregated classification	m³/ha
Woodland[a]	71
Bush[b]	32
Shrub[c]	14
Saltpan	0

[a] Woodland includes the ETC categories of "Dense Woodland" and "Open Woodland."

[b] Bush includes the ETC category "Woodland and Bushland."

[c] Shrub includes ETC categories of "Bushland with Scrubby Woodland and Woody Shrubland," "Shrubland and Bushy Shrubland," and "Hill Shrubland and Woodland."

Cape Verde

Vegetation Map. *Source:* "Republic of Cape Verde, Forest Resources." 1984. IBRD # 18134R. Date: September 1984. Scale: 1:1 million. Projection: Unknown.

Standing Stock Density Derivations. All wooded areas in Cape Verde are government-owned and -operated plantations. The areal extent of the two types of plantations (high-altitude and low-altitude) were mapped and calculated; however, volume estimates were not completed. Cape Verde was eliminated from the woodfuels supply analysis because of the unusual nature of the biomass, and stand density estimates were not further pursued.

Haiti

Vegetation Map. *Source:* Principal Forest Resources of Haiti. IBRD #22676 in "Haiti Forestry and Environmental Protection Project." Staff Appraisal Report, Report #9307-HA, September 3, 1991. Date: Published 1990; data collected several years earlier. Scale: 1:1 million. Projection: Unknown. Notes: The map is dated 1990, but much of the information is 10 or more years old, and the forest resource may be half of what is depicted (Goetz 1991). To account for this, all demographic and economic data were back-extrapolated to 1980.

Standing Stock Density Derivations. Biomass density information was derived from a study completed for the Haitian Ministry of Economy and Finance by a private consulting group (BDPA 1989). Two assumptions were made about the information in this report.

- Density figures represent stemwood only and do not include crown and understory.
- Coffee plantation figures describe a plantation consisting of short woody plants beneath a partial canopy. Fuelwood figures include trimmings of branches from the coffee crop and all wood available from the short woody plants.

The following ratios of total biomass to stemwood were applied to determine the adjusted density values below:

BDPA classification	Forest category	Total biomass/stemwood	Adjusted m³/ha
Dense Pine Forest	Production Forest	2.1	300
Open Pine Forest	Open Forest	2.8	55
Broadleaf Forest	Production Forest	2.1	300
Degraded Broadleaf	Degraded Forest	NA	80

Note: NA = not available.

Indonesia

Vegetation Map. *Source:* Ministry of Forestry, Indonesia. Date: 1985. Scale: 1:1 million. Projection: Unknown. Notes: The vegetation map was not used as the base map in Indonesia. *The Times Atlas of the World* map of Java served as the base map and therefore there was not perfect alignment with the vegetation map; error in forest placement may have occurred as a result, but area calculations of each forest parcel are still accurate.

Standing Stock Density Derivations. The biomass assessment methodology established in prior work on this project was followed (Evron 1990). By this methodology, 17 forest categories were aggregated into three categories that coincided with the classification system of the FAO GEMS study. Adjustments for crown and understory were then applied.

Aggregated classification	FAO forest category	Total biomass/ stemwood	Adjusted m^3/ha
Dense Forest	Protection	1.5	252
Open Woodland	Production	2.1	193
Degraded Woodland	Degraded	0.52	88

Mauritania

Vegetation Map. *Source:* Millington et al. 1992. Date: 1985–1987 imagery from the AVHRR sensor of the NOAA-7 satellite was the basis for this map. Scale: 1:5 million. Projection: Unknown. Notes: The vegetation map was used as the base map, with scale corrections made using the *Times Atlas of the World* (1990) map of Mauritania.

Standing Stock Density Derivations. Biomass density values for Mauritania were taken from the Millington et al. (1994) report, which conforms to the format of the woodfuels supply assessment and therefore did not need to be adjusted.

Vegetation Classification	m^3/ha
Desert	0
Wooded Grassland	3
Bush and Thicket	32

Philippines

Vegetation Map. *Source:* Swedish Space Corporation: "Philippines Land Use Map, Natural Conditions of the Philippines." Date: 1988, based on SPOT multispectral satellite images taken between 1987 and 1988, on a scale of 1:100,000. Ground truthing was performed May–June 1987. Scale: 1:2 million. Projection: Transversal Mercator Projection

Standing Stock Density Derivations. The FAO document was used as the basis for all density estimates, except for grasslands and plantations. Expert opinion on wooded grasslands and plantations were provided by World Bank staff (Openshaw 1992).

Swedish space classification	Forest category	Adjustment factor	m³/ha
Closed Broadleaf Forest	Production/Protection[b]	345	
Open Broadleaf Forest	Open Forest[c]	2.8	255
Pine Forest	Production Forest	2.1	200
Mangrove	NA	NA	95
Wooded Grassland	NA	NA	8
Plantations[a]	NA	NA	84

[a] Plantations in the mapped regions of Thailand are predominantly coconut, with some incidence of rubber and coffee. The density value is based on the assumption that the average MAI for this collective group of plantations is 5 metric tons/ha/year, MAI is approximately 6 to 10 percent of the growing stock, there are roughly 1.4 m3/ha per 1.0 metric ton/ha, and reflects all aboveground biomass.

[b] Based on an aerial weighted average of the two categories.

[c] The biomass density of the Open Broadleaf Forest was not given for Thailand in the GEMS study. Thai value was derived from the ratio of Open Forest to Protection Forest in several other Asian countries. The Open Forest was on average 30 percent of the Undisturbed Closed Forest.

Note: NA = Not applicable

Thailand

Vegetation Map. *Source:* "Forest Types Map," prepared by the Royal Forestry Department, Forest Management Division, Remote Sensing and Mapping Sub-Division (Bangkok). Date: 1983. Based on the interpretation of aerial photographs taken during 1972–1977, and LANDSAT images taken in 1978. Scale: 1:1 million. Projection: Unknown

Standing Stock Density Derivations. The FAO document was referenced for all biomass density estimates except for Rubber Plantations. Mangrove figures were not reported by FAO for Thailand, so this information was estimated from mangrove densities in other Asian countries and adjusted with World Bank expert opinion (Ryan 1992).

Royal forestry classification	FAO forest category	Total biomass / stemwood	Adjusted m³/ha
Tropical Evergreen	Closed Forest	1.5	120
Mixed Deciduous Forest	Closed Forest	1.5	120
Dry Dipterocarp	Open Forest	2.8	78
Pine	Closed Forest	1.5	90
Rubber Plantation[a]	Plantation	NA	120
Mangrove	NA	NA	100

[a] Estimate provided by World Bank expert (Openshaw 1992) and reflects all aboveground biomass.
Note: NA = not available.

Zimbabwe

Vegetation Map. *Source:* ETC-Foundation. Date: 1987; data collected between December 1985 and March 1987. Scale: 1:5 million. Projection: Unknown. Notes: This map was based on interpretation of satellite imagery from the NOAA-7 meteorological satellite, with a resolution of 8 km. Remotely sensed data were verified with field data, maps, and other secondary sources. The country border was digitized from the *Times Atlas of the World* (1990).

Standing Stock Density Derivations. Standing biomass density estimates were taken completely from the ETC report on the SADCC member states. These values include all aboveground biomass and do not make reductions for merchantable timber, and they therefore do not need to be adjusted for crown and understory. The ETC values of metric tons per hectare were converted into cubic meters per hectare at a rate of 1.4 cubic meter per hectare per 1 metric ton per hectare. Because of the complexity of the map, similar vegetation types were aggregated, and biomass density values were calculated as an aerial weighted average as follows:

Aggregated classification	m³/ha
Dense Savannah Woodland	103
Open Montane Woodland[a]	62
Dry Bushy Savannah	33
Wooded Grassland	16

[a] This category includes four ETC categories: "Open Savannah and Baikiaea Woodland and Montane Vegetation," "Seasonal Savannah Woodland," "Dry Savannah Woodland," and "Mopane Woodland and Escarpment Thicket." Because of the aggregation of these vegetation classifications, skewing of the biomass distribution may have occurred.

Notes

1. An exception is a classic study that conducted a cross-country comparison of conditions and trends in South Asia (Leach 1986, 1987), offering comparative insight about urban energy transitions in that region. An insightful cross-city approach is also offered by Sathaye and Tyler (1991).

2. Among others, these studies include World Bank 1988, 1989, 1990a, 1990b, 1990c, 1990d, 1991a, 1991b, 1992, 1993, 1996a, 1999.

3. An annex at the end of the book describes the data sources and estimation methodology for determining the spatial distribution of biomass resources in the periurban regions studied.

4. This conclusion may not hold in some urban areas where modern fuels are available but quantity is rationed. This issue is discussed in more detail in Chapter 4.

5. Fuelwood harvest and distribution costs should be inversely related to the level of standing biomass stock and its proximity to urban consumer markets. This supply-side linkage suggests that, holding the demand side of the market and other factors constant, households should have a relatively large incentive to consume biomass fuels if they live in proximity to abundant biomass resources.

6. Electricity access in Pakistan is more than 80 percent (Eiswerth et al. 1998).

7. We do not have the information to adjust fuel costs in this study for the other factors raising its costs to low-income consumers (e.g., the time opportunity cost to woodfuel collection and/or utilization, and nonmonetized health effects from indoor air quality problems).

8. Ultimately, we did not include the cities of Cape Verde in the deforestation analysis, due to the fact that the periurban areas in question were too dry to contain biomass.

9. This chapter presents the results of a more detailed report entitled *India: Household Energy Strategies for Urban India: the Case of Hyderabad* (World Bank 1999), which was produced as part of the World Bank's Energy Sector

Management Assistance Program. The study is unique because researchers were able to compare recent energy use to a similar study conducted in 1981–1982.

10. The price of electricity was raised in 1996. It now ranges from Rs 0.80/kWh, for consumers consuming less than 50 kWh per month, to Rs 2.6/kWh for customers consuming more than 400 kWh per month.

11. The estimated demand for fuelwood and charcoal given here excludes the use of these fuels by small-scale industry. A cursory survey was done in some fuelwood-using industries such as edible oil extraction mills, textile plants (silk and tery-lene), footwear makers, utensil manufacturers, and chemical works (in addition, a negligible quantity of charcoal may be used by blacksmiths). Many owners were reluctant to give figures of consumption, because they were using the wood to supplement or substitute for (subsidized) coal, for which they had been allocated a quota. The general consensus was that although wood was more expensive than coal, the supply was reliable. Also, it is possible that some of the coal was resold at market prices. The wood-using industries, such as sawmills, also used wood waste as a boiler fuel to kiln dry sawn wood. The 1994 estimated consumption of fuelwood by these small-scale industries is about 45,000 metric tons. (Alam 1996). In comparison, the 1982 demand in this small industries sector may have been about 24,000 metric tons of fuelwood.

12. These policies are all sector specific. Governments could implement other policies, such as developing labor-intensive local industry or instituting social programs to reduce poverty or improve public health, that could help reduce externalities in urban energy markets or improve the welfare of lower-income consumers. An analysis of this broader set of policy options, however, is outside the scope of the present study.

13. Biomass density was calculated from Borries' MAI figures, assuming that MAI is roughly 2 to 4 percent of the stand density in these natural mixed-age stands. However, the values were significantly higher than all values presented in the FAO GEMS study: calculated values ranged from 350 to 700 m³/ha; FAO figures were between 100 and 250 m³/ha. For project consistency and regional comparison, FAO values were adopted.

References

Aburas, R., and J.-W. Fromme. 1991. Household Energy Demand in Jordan. *Energy Policy* July/August 19(6): 589–595.

Adegbulugbe, A.O., and J.F.K. Akinbami. 1995. Urban Household Energy Use Patterns in Nigeria. *Natural Resources Forum* 19(2): 125–133.

Agarwal, B. 1986. *Cold Hearths and Barren Slopes.* London: Zed Press.

Alam, Manzoor, Joy Dunkerley, K.N. Gopi, William Ramsay, and E. Davis. 1985a. *Fuelwood in Urban Markets: A Case Study of Hyderabad.* New Delhi: Concept Publishing Co.

Alam, Manzoor, Joy Dunkerley, and A.K. Reddy. 1985b. Fuelwood Use in the Cities of the Developing World: Two Case Studies from India. *Natural Resources Forum* 9(3): 205–213.

Alam, Manzoor, Jayant Sathaye, and Douglas Barnes. 1998. Urban Household Energy Use in India: Efficiency and Policy Implications. *Energy Policy* 26(11): 885–891.

Allen, Julia, and Douglas Barnes. 1985. The Causes of Deforestation in Developing Countries. *Annals of the Association of American Geographers* 75(2): 163–184.

Arnold, J.E.M. 1979. Wood Energy and Rural Communities. *Natural Resources Forum* 3: 229–252.

Barnes, Douglas F. 1990. Population Growth, Wood Fuels, and Resource Problems in Sub-Saharan Africa. In *Population Growth and Reproductions in Sub-Saharan Africa,* edited by Rudolfo Bulatao and George Ascadi. Baltimore: World Bank and John's Hopkins University Press.

Barnes, Douglas F., and Willem Floor. 1996. Rural Energy and Developing Countries: A Challenge for Economic Development. *Annual Review of Energy and Environment* 21: 497–530.

———. 1999. Biomass Energy and the Poor in the Developing World. *Journal of International Affairs* 53(1):237–262.

Barnes, Douglas F., and Liu Qian. 1992. Urban Interfuel Substitution, Energy Use, and Equity in Developing Countries: Some Preliminary Findings. In *International Issues in Energy Policy, Development, and Economics*, edited by James Dorian and Fereidun Fesharaki. Boulder: Westview Press, 163–182.

Bensel, Terrence. 1995. Rural Woodfuel Production for Urban Markets: Problems and Opportunities in the Cebu Province, Philippines. *Pacific and Asian Journal of Energy* (New Delhi, India) 5(June): 9–28.

Bhatia, Ramesh. 1985. Energy Pricing in Developing Countries: Role of Prices in Investment Allocation and Consumer Choices. In *Criteria for Energy Pricing Policy*, edited by C.M. Sidday. London: Graham and Trotman.

———. 1988. Energy Pricing and Household Energy Consumption in India. *Energy Journal* (Special South and Southeast Asia Pricing Issue) 9: 71–105.

Boberg, Jill. 1993. Competition in Tanzania Woodfuel Markets. *Energy Policy* 21(5): 474–490.

Bowonder, B., S.S.R. Prasad, and K. Raghuram. 1987a. Fuelwood Use in Urban Centres: A Case Study of Hyderabad. *Natural Resources Forum* 11(2): 189–195.

Bowonder, B., S.S.R. Prasad, and N.V.M. Unni. 1987b. Deforestation around Urban Centers in India. *Environmental Conservation* 4(1/Spring): 23–28.

———. 1988. Dynamics of Fuelwood Prices in India: Policy Implications. *World Development* 16(10): 1213–1229.

Brouwer I.D., J.C. Hoorweg, and M.J. Van Liere. 1997. When Households Run Out of Fuel: Responses of Rural Households to Decreasing Fuelwood Availability. Ntcheu District, Malawi. *World Development* 25(2): 255–266.

Brown, Sandra, and Ariel Lugo. 1984. Biomass of Tropical Forests: A New Estimate Based on Forest Volumes. *Science* 223(23/March): 1290–1293.

Cernea, Michael M. 1981. Land Tenure Systems and Social Implications of Forestry Development Programs. World Bank Staff Working Paper 452. Washington, DC: World Bank.

Chauvin, Henri. 1981. When an African City Runs Out of Fuel. *Unasylva* 33(133): 11–21.

Cline-Cole, R.A., H.A.C. Main, and J.E. Nichol. 1990. Fuelwood Consumption, Population Dynamics, and Deforestation in Africa. *World Development* 18: 513–527.

Darrat, A.F. 1999. Are Financial Deepening and Economic Growth Causally Related? Another Look at the Evidence. *International Economic Journal* 13(3): 19–35.

Dasgupta, Partha. 1998. The Economics of Poverty in Poor Countries. *Scandinavian Journal of Economics* 100(1): 41–68.

Dasgupta, S., B. Laplante, H. Wang, and D. Wheeler. 2002. Confronting the Environmental Kuznets Curve. *Journal of Economic Perspectives* 16(1): 147–168.

de Martino Jannuzzi, G., and L. Schipper. 1991. The Structure of Electricity Demand in the Brazilian Household Sector. *Energy Policy* November 19(11): 879–891.

Dewees, P.A. 1989. The Woodfuel Crisis Reconsidered: Observations on the Dynamics and Abundance and Scarcity. *World Development* 17: 1159–1172.

———. 1995. Forestry Policy and Woodfuel Markets in Malawi. *Natural Resources Forum* 19(2): 143–152.

Eckholm, E. 1975. *The Other Energy Crisis: Firewood*. Worldwatch Paper 1. Washington, DC: Worldwatch Institute.

Eiswerth, M.E., K.W. Abendroth, R.E. Ciliano, A. Ouerghi, and M.T. Ozog. 1998. Residential Electricity Use and the Potential Impacts of Energy Efficiency Options in Pakistan. *Energy Policy* 26(4): 307–315.

Estache, A., A. Gomez-Lobo, and D. Leipziger. 2001. Utilities Privatization and the Poor: Lessons and Evidence from Latin America. *World Development* 29(7): 1179–1198.

Fitzgerald, Kevin, Douglas Barnes, and Gordon McGranahan. 1990. *Interfuel Substitution and Changes in the Way Households Use Energy: The Case of Cooking and Lighting Behavior in Urban Java.* Industry and Energy Department Working Paper, Energy Series Paper 29. Washington, DC: World Bank.

Foley, Gerald. 1987. Exaggerating the Sahelian Fuelwood Problem? *Ambio* 16: 67–71.

Foley, Gerald, Willem Floor, Gerard Madon, Elahadji Mumamane Lawali, Pierre Montague, and Kiri Tounao. 1997. *The Niger Household Energy Project: Promoting Rural Fuelwood Markets and Village Management of Natural Woodlands.* World Bank Technical Paper 362. Washington, DC: World Bank.

Hartwick, J.M. 1992. Deforestation and National Accounting. Environmental and *Resource Economics* 2: 513–521.

Hosier, Richard H. 1993. Urban Energy Systems in Tanzania: A Tale of Three Cities. *Energy Policy* 21(5): 510–523.

Hosier, Richard H., and Jeffrey Dowd. 1988. Household Fuel Choice in Zimbabwe: An Empirical Test of the Energy Ladder Hypothesis. *Resources and Energy* 9: 347–361.

Hosier, Richard H., and W. Kipondya. 1993. Urban Household Energy Use in Tanzania: Prices, Substitutes and Poverty. *Energy Policy* 21(5): 454–473.

Hughes, Gordon, Kseniya Lvovsky, and Meghan Dunleavy. 2001. *Environmental Health in India: Priorities in Andhra Pradesh.* Washington, DC: World Bank.

Hyde, William, and Yuan Seve. 1993. The Economic Role of Wood Products in Tropical Deforestation: The Severe Example of Malawi. *Forest Ecology and Management* 57(2): 283–300.

Hymen, Eric L. 1983. Analysis of the Woodfuels Market: A Survey of Fuelwood Sellers and Charcoal Makers in the Province of Ilocos Norte, Philippines. *Biomass* 3: 167–197.

———. 1985. Demand for Woodfuels by Households in the Province of Ilocos Norte, Philippines. *Energy Policy* 8: 581–591.

Kammen, E. 2001. Quantifying the Effects of Exposure to Indoor Air Pollution from Biomass Combustion on Acute Respiratory Infections in Developing Countries. *Environmental Health Perspectives* 109(5): 481–488.

Kumar, M.S. (ed.). 1987. *Energy Pricing Policies in Developing Countries: Theory and Empirical Evidence.* New York: United Nations Development Program and Economic and Social Commission for Asia and the Pacific.

Kumar, Shubh K., and David Hotchkiss. 1988. *Consequences of Deforestation for Women's Time Allocation, Agricultural Production, and Nutrition in Hill Areas of Nepal.* Washington, DC: International Food Policy Research Institute

Lan, Qing, Robert Chapman, Dina Shreinemachers, Linwei Tian, and Xingzhou He. 2002. Household Stove Improvement and Risk of Lung Cancer in Xuanwei, China. *Journal of the National Cancer Institute* 94(11): 826–835.

Leach, Gerald. 1986. *Household Energy in South Asia.* London: International Institute for Environment and Development.

————. 1987. Household Energy in South Asia. *Biomass* 12: 155–184.

————. 1993. The Energy Transition. *Energy Policy* 21(2): 116–123.

Lewis, Laurence, and William Coffey. 1985. The Continuing Deforestation of Haiti. *Ambio* 14(3): 158–160.

Malawi Ministry of Forestry and Natural Resources. 1984. *Malawi: Urban Energy Survey*. Lilongwe: Energy Studies Unit.

Mercer, E., and J. Soussan. 1990. Fuelwood: An Analysis of Problems and Solutions for Less Developed Countries. *The World Bank Policy Review*. Washington, DC: World Bank.

Millington, A.C., R.W. Critchley, T.D. Douglas, and P. Ryan. 1990 in text, p. 14. *Estimating Woody Biomass in Sub-Saharan Africa: The Application of Remote Sensing for Woody Biomass Assessment and Mapping*. Washington, DC: World Bank.

Milukas, M.V. 1986. Energy Flow in a Secondary City: A Case Study of Nakuru, Kenya. Ph.D. thesis. University of California, Berkeley.

Nair, K.N.S., and J.G. Krishnayya. 1985. *Energy Consumption by Income Groups in Urban Areas of India*. World Employment Programme Research Working Paper, Technology and Employment Programme. Geneva: International Labour Organization.

Nieuwenhout, F., P. Van de Rijt, and E. Wiggelinkhuizen. 1998. *Rural Lighting Services*. Paper prepared for the World Bank by The Netherlands Energy Research Foundation, The Netherlands. Washington, DC: World Bank.

Openshaw, Keith. 1986. Methods of Collecting Biomass Supply Statistics. Paper prepared for the Commonwealth Secretariat Biomass Resources Assessment Workshop, Mauritius. December 1986, Nairobi, Kenya.

————. June 1992. Personal communication with the authors.

Panayotou, T. 2003. *Economic Growth and the Environment*. Paper presented at the spring seminar of the United Nations Economic Commission for Europe. Geneva, Switzerland, March 3, 2003.

Peskin, Henry, Willem Floor, and Douglas Barnes. 1992. *Accounting for Traditional Fuel Production: The Household Energy Sector and its Implications for the Development Process*. Industry and Energy Department Working Paper, Energy Series Paper No. 49. Washington, DC: World Bank.

Reddy, A.K., and B.S. Reddy. 1983. Energy in a Stratified Society: Case Study of Firewood in Bangalore. *Economic and Political Weekly* 17(1): 1757–1770.

————. 1984. Substitution of Energy Carriers for Cooking in Bangalore. *Energy* 19: 561–572.

Roberts, T.J., and P.E. Grimes. 1997. Carbon Intensity and Economic Development 1962–1991: A Brief Exploration of the Environmental Kuznets Curve. *World Development* 25(2): 191–198.

Ryan, Paul. June 1992. Personal communication with the authors. *Basic Methodology for Adjusting Merchantable Timber Figures to Include Crown and Understory*.

Sathaye, Jayant, and Steven Meyers. 1985. Energy Use in Cities of the Developing Countries. *Annual Review of Energy* 10:109–133.

Sathaye, Jayant, and Stephen Tyler. 1991. Transition in Household Energy Use in Urban China, India, the Philippines, Thailand, and Hong Kong. *Annual Review of Energy and Environment* 16: 295–335.

Sharma, Rishi, and Bhatia Ramesh. 1986. *Basic Energy Needs of the Low-Income Groups in India: Analysis of Energy Policies and Programs*. Report for the Regional Energy Development Program, ILO and ARTEP. New Delhi, India: International Labor Organization

Smith, Kirk. 1993. Fuel Combustion, Air Pollution Exposure, and Health: The Situation in Developing Countries. *Annual Review of Energy and Environment* 18: 529–566.

———. 2002. Indoor Air Pollution in Developing Countries: Recommendations for Research. *Indoor Air* 12: 198–207.

———. 2003. Indoor Air Pollution and Acute Respiratory Infections. *Indian Pediatrics* 40: 815–819.

Smith, Kirk, and Sumi Mehta. 2003. The Burden of Disease from Indoor Air Pollution in Developing Countries: Comparison of Estimates. *International Journal of Hygiene and Environmental Health* 206: 279–289.

Smith, Kirk, Jonathan M. Samet, Isabelle Romeiu, and Nigel Bruce. 2000. Indoor Air Pollution in Developing Countries and Acute Lower Respiratory Infections in Children. *Thorax* 55: 518–532.

Stern, D.I., and M.S. Commons. 2001. Is There an Environmental Kuznets Curve for Sulfur? *Journal of Environmental Economics and Management* 41: 162–178.

Stevenson, Glenn. 1989. The Production, Distribution, and Consumption of Fuelwood in Haiti. *Journal of Developing Areas* 24: 59–76.

Tibesar, A., and R. White. 1990. Pricing Policy and Household Energy Use in Dakar, Senegal. *The Journal of Developing Areas* 25: 33–48.

Tuan, Nguyen Anh, and Thierry Lefevre. 1996. Analysis of Household Energy Demand in Vietnam. *Energy Policy* 24(12): 1089–1099.

van der Plas, Robert, and A. van de Graaff. 1988. *A Comparison of Lamps for Domestic Lighting in Developing Countries*. Industry and Energy Department Working Paper, Energy Series 6. Washington, DC: World Bank.

Wallmo, K., and S. Jacobson. 1998. A Social and Environmental Evaluation of Fuel-Efficient Cook-Stoves and Conservation in Uganda. *Environmental Conservation* 25(2): 99–108.

Wang, Xiaohua, and Feng Zhenmin. 2001. Rural Household Energy Consumption with Economic Development in China: Stages and Characteristic Indices. *Energy Policy* 29: 1391–1397.

Westley, Glenn D. 1992. *New Directions in Econometric Modeling of Energy Demand: With Applications to Latin America*. Washington, DC: Inter-American Development Bank.

World Bank. 1988. *Niger: Household Energy Conservation and Substitution*. Report of the Joint UNDP/World Bank Energy Sector Management Assistance Programme, January. Washington, DC: World Bank.

———. 1989. *Senegal: Urban Household Energy Strategy*. Report of the Joint UNDP/World Bank Energy Sector Management Assistance Programme, June. Washington, DC: World Bank.

———. 1990a. *Cap Vert: Strategies Energetiques dans le Secteur Residentiel Enquetes Consommateurs*. Report of the Joint UNDP/World Bank Energy Sector Management Assistance Programme, October. Washington, DC: World Bank.

————. 1990b. *Indonesia: Urban Household Energy Strategy Study.* Main Report, Report 107A/90. Washington, DC: World Bank.

————. 1990c. *Mauritania: Elements of Household Energy Strategy.* Report 123/90. Washington, DC: World Bank.

————. 1990d. *Zambia: Urban Household Energy Strategy.* Report 121/90. Report of the Joint UNDP/World Bank Energy Sector Management Assistance Programme. Washington, DC: World Bank.

————. 1991a. *Burkina Faso: Urban Household Energy Strategy.* Report 134/91. Washington, DC: World Bank.

————. 1991b. *Haiti: Household Energy Strategy.* Report 143/91. Washington, DC: World Bank.

————. 1992. *Republic of Mali: Household Energy Strategy.* Report 147/92. Washington, DC: World Bank.

————. 1993. *Lao PDR: Urban Energy Demand Assessment.* Joint UNDP/ESMAP Report 154/93. Washington, DC: World Bank.

————. 1996a. *China: Energy for Rural Development in China: An Assessment Based on a Joint Chinese/ESMAP Study of Six Counties.* Joint UNDP/ESMAP Report 183/96. Washington, DC: World Bank.

————. 1996b. *Rural Energy and Development: Improving Energy Supplies for Two Billion People*: Development in Practice Series. Washington, DC: World Bank.

————. 1998. *World Development Report 1998: Development and the Environment.* New York: Oxford University Press.

————. 1999. *India: Household Energy Strategies for Urban India: The Case of Hyderabad.* Joint UNDP/ESMAP Report 214/99. Washington, DC: World Bank.

————. 2002. *Rural Electrification and Development in the Philippines: Measuring the Social and Economic Benefits.* Joint UNDP/ESMAP Report. Washington, DC: World Bank.

Index